IMPERFECT ORACLE

*In this place I am minded to build a glorious temple to be an oracle for men, and here they will always bring perfect hecatombs, both they who dwell in Peloponnesus and the men of Europe and from all the wave-washed isles, coming to question me. And I will deliver to them all counsel that cannot fail, answering them in my rich temple.*

HOMER, *HYMN TO PYTHIAN APOLLO*, TRANS. EVELYN WHITE

# IMPERFECT ORACLE

## THE EPISTEMIC AND MORAL AUTHORITY
## OF SCIENCE

*Theodore L. Brown*

THE PENNSYLVANIA STATE UNIVERSITY PRESS
UNIVERSITY PARK, PENNSYLVANIA

Library of Congress Cataloging-in-Publication Data

Brown, Theodore L. (Theodore Lawrence), 1928–
Imperfect oracle : the epistemic and moral authority of science/Theodore L. Brown.
p.        cm.
Summary: "Explores the relationships between science and other societal sectors, notably law, religion, government and public culture, in terms of the concepts of expert and moral authority"—Provided by publisher.
Includes bibliographical references and index.
ISBN 978-0-271-03535-2 (cloth : alk. paper)
1. Science—Moral and ethical aspects.
2. Science—Social aspects.
I. Title.

Q180.55.M67B76 2009
174′.95—dc22
2009000901

It is the policy of The Pennsylvania State University Press
to use acid-free paper. This book is printed on Natures Natural,
containing 50% post-consumer waste, and meets the minimum
requirements of American National Standard for Information Sciences—
Permanence of Paper for Printed Library Material, ANSI Z39.48–1992.

*for* AUDREY

# CONTENTS

# PREFACE

We all know that science, whether we love it, abide it, or even detest it, is here to stay; we also know that science and technology, in important senses, define modern culture. They are the agents on which much else depends, whether it's food and fresh water for the world's billions, or new fabrics being shown on the fashion runways of Paris and New York. From matters of life and death to trivial pursuits, contemporary life hinges on science and technology. There is no turning back the clock on scientific and technological "progress"; even the most resolutely organic farm commune finds itself partaking of scientific knowledge.

It is this very omnipresence that gives people the willies, as indeed it should. If the world holds itself in thrall to all the instrumental offerings of science and technology with little regard for their larger implications—for what they mean for our destiny as a species or for our moral obligations to one another and to the world in which we live—human civilization is not long for this planet. There is really not a great deal of time in which to figure this out; events are overtaking us. Each day the global human population increases by more than two hundred thousand. Each day about seventy-seven million metric tons of carbon dioxide are added to the atmosphere as a result of human activity. Each day, on average, the sea level rises just a little.

We seem to have overrun ourselves as a species; we've been very clever in dealing with the physical world, using our evolutionary, biological inheritance to such good effect that we can do quite amazing things. But we still carry with us cognitive predilections that served humanity well on its way up the evolutionary slopes, but which may now contribute to our undoing. A rational approach to study of the physical world that employs the methodologies of scientific research has provided humanity with a cornucopia of beneficial products. Ironically, some of these have consequences that can be disastrous over time if not properly dealt with. Science has much to offer in addressing the complex social, economic, and political problems confronting society, but its voice is only one of many. There is no doubt that science exercises influence, but it often falls short of what it might attain. So it's worth asking why, apart from simply delivering a great many products, science does not have a stronger voice in shaping the culture of society.

I've chosen to analyze this and important related questions by using the concept of "authority" as a kind of lens through which to view science's interactions with the larger society. I've learned in the course of writing the book that "authority" is a multifaceted idea. Part of my task has been to identify the kinds of authority exercised by science and to show how scientific authority stands in relationship to the authorities that characterize other social sectors. I hope that this book will yield a clearer sense both of how science is blended into the fabric of society as a whole and of how and why it so often fails to make the contributions to public discourse of which it is capable.

I have organized the book into two parts. In the first I consider authority generally: a classification in terms of the kinds of authority that exist, and their origins in social institutions. I then deal with historical matters: how science came to have authority, and the contests with other societal sectors through which that authority was won. I provide a brief history of science in America and follow that with a look at the place of science in contemporary society. The emphasis throughout is on the growth of science's capacity to exercise authority, and on the factors that either limited or enhanced that capacity. Thus the historical narratives and discussion of science in contemporary society are focused; they are not meant to be fully fleshed-out treatments.

In the second half of the book I turn to examining the roles of science in several key social sectors: law, religion, government, and the public. Any one of the chapters in this part of the book could be greatly expanded beyond my coverage. My goal has been to explain the ways in which authority figures into the relationships between science and each of these sectors. In some instances the relationship is adversarial, but certainly not always. I have chosen topical areas within each of these sectors as vehicles for demonstrating how science's authority and autonomy are exercised and sometimes limited. Once again, none of these chapters is meant to be broadly inclusive of all that can be said. The particular topics chosen are meant to be exemplary of the interactions between science and the particular social sector. Nor have I attempted to exercise a great deal of moral authority; the book is more descriptive than prescriptive. Nevertheless, there are many places in which I take advantage of the bully pulpit of authorship to argue on one side or another of a controversial matter.

I am indebted to many who have been kind enough to engage in extended conversations, read and comment on chapters, or guide me to appropriate sources, particularly Peter Beak, Audrey Brown, Jennifer Brown, Mark Brown, Mary Coffield, Richard Burkhardt, Ira Carmen, Simon Cole, Steve Gimbel, Robert Jones, Jay Labinger, Michael Lynch, Robert Switzer, and two anonymous reviewers. Loren Zech was especially helpful in the early going, helping me locate needed resource materials and providing feedback on my early drafts. I am much obliged to the friendly people at Pennsylvania State University Press, notably Sandy Thatcher, director of the Press, Kathryn Yahner, Jennifer Norton, Patricia Mitchell, Steve Kress, and Laura Reed-Morrisson. I owe a debt of gratitude to Nicholas Taylor for his excellent editorial work. I am grateful also to Pierre Wiltzius, former director of the Arnold and Mabel Beckman Institute for Advanced Science and Technology at the University of Illinois, Urbana-Champaign, for providing funds for a research assistant in the early stage of the project.

Finally, I thank my spouse, Audrey Brown, for her insights and for indulging my obsession with the project over such a long time.

# ABBREVIATIONS

| | |
|---|---|
| AAAS | American Association for the Advancement of Science |
| ABA | American Bar Association |
| AMA | American Medical Association |
| APS | American Physical Society |
| IPCC | Intergovernmental Panel on Climate Change |
| NAS | National Academy of Sciences |
| NIH | National Institutes of Health |
| NRC | National Research Council |
| NSF | National Science Foundation |

The Innocence Project at the Benjamin N. Cardozo School of Law at Yeshiva University, founded by Barry C. Scheck and Peter J. Neufeld in 1992, is a nonprofit legal clinic and criminal justice resource center devoted to exonerating those wrongfully convicted. At its Web site one can find a color photo of Eddie Joe Lloyd, an African American man of early middle age. Eddie's skin is the color of caramel; his hair, worn very short, appears to be gray. He is dressed in a white shirt, tie, and dark suit. He's a nice-looking man, wearing what seems to be a rather forced smile.

Eddie Joe Lloyd was convicted in 1984 of the brutal murder of a sixteen-year-old girl in Detroit, Michigan. While in a hospital as a mental patient, Lloyd wrote to the police suggesting how to solve various open murder cases. The police interviewed him in the hospital, and in the course of their interrogations, they allowed him to believe that by confessing to the murder of the sixteen-year-old and being arrested, he would help them "smoke out" the real perpetrator. They provided Lloyd with details of the crime, including such matters as what clothing and jewelry she was wearing. Lloyd then signed a written confession and gave a tape-recorded statement. He was put on trial and convicted of the murder after less than an hour of jury deliberation. Against the advice of his defense attorney, he refused to plead mental incompetence. The sentencing judge, Leonard Townsend, complained that he could only sentence Lloyd to life imprisonment rather than hanging; Michigan's repeal of the death penalty prevented the judge from imposing the sentence he thought appropriate.

After exhausting all his appeal possibilities while in prison, Lloyd contacted the Innocence Project. Most of the Project's clients are poor and largely forgotten, and they have used up all their normal avenues of legal appeal. In Lloyd's case, Project students searched for years to uncover materials that could form the basis of a convincing DNA test. Eventually, DNA profiles from several different items of evidence from the crime scene matched one another and excluded Eddie Joe Lloyd. The Innocence Project was joined by the Wayne County Prosecuting Attorney's office and the Detroit Police Department in filing a motion to vacate Lloyd's conviction. In 2002, after serving seventeen years in prison, Lloyd became the 110th person in the United States to be exonerated by post-conviction DNA testing.

Faculty and students at the Cardozo School of Law, the Northwestern University School of Law, and institutions elsewhere throughout the nation have helped bring about more than 216 post-conviction exonerations in the United States. (You can learn the latest number at the Innocence Project Web site: http://www.innocenceproject.org/Content/351.php.) In the process, they have prompted a reevaluation of the death penalty in light of the unacceptably high incidence of wrongful convictions. It is a great story of injustices righted; it is also quite relevant to the theme of this book. Consider this: any jurist or judge who votes to consign an accused to execution or life imprisonment must surely have a strong, virtually unshakable belief in that person's guilt. Yet a laboratory scientist, armed with a few projection slides, can overturn the aggregate of those beliefs in a few minutes of presentation. How can tests run on biological samples taken from the crime scene, the victims, and the accused have the power to change minds in a matter on which belief must have been fixed so powerfully?

It is as though an unimpeachable witness had come forward to attest that the accused was somewhere else at the time of the crime, perhaps even providing corroborating evidence of his presence there. But what does it take for a witness to be unimpeachable? She must have a character widely recognized as beyond reproach, a proven record of honesty, trustworthiness, and sound judgment. Further, the witness must have no direct interest in the outcome, other than to see justice done. The mother of the accused will not do, but someone in contact with the accused at the time of the crime—such as a bank loan officer, someone interviewing the accused for a job, or a police officer who might have been handing out a traffic ticket to the accused at a distant location—would all potentially qualify. In a similar way, DNA testing serves as the unimpeachable witness.

One can hope to judge the qualities that might make a person an unimpeachable witness, and put her relationship to the accused into perspective.

But what sort of informed judgment can someone who is nonexpert in the science that underlies DNA testing form about the assertion that the laboratory results show—not just hint at, but *show unequivocally*—that the DNA taken from the crime scene and the accused are not the same stuff? For nearly everyone, including the judge or jury who must decide on the evidence, the answer is, virtually none.

To be sure, the slides show clearly that the little bars of darkness derived from the samples do not line up with those derived from the accused. Of course, there is a discourse on how the samples were obtained and treated to get to the slide materials. But people in general really do not know what DNA is at a molecular level, nor do they have any clear idea of the theory of DNA, how it comes to be in the materials gathered, or how the samples give rise to the analyzed pictures. Society simply takes the word of scientists for all that—it accepts science's authority.

This book is about the authority of science, and about other matters related to that core idea, such as autonomy and moral authority. The story with which I have begun is a simple illustration of science exercising authority. In this case, it is acting as our hypothetical unimpeachable witness. What science declares to be the case, just as what an unimpeachable witness testifies, is taken to be true and relevant to the matter at hand. But our example does nothing to clarify *how* science can exert such an authority. Why is it that society accepts a conclusion based on a laboratory test even though the science that underlies that conclusion is clouded in mystery to all but a few?

Acceptance does not happen automatically; it has to be somehow earned, not just for science in general, but in any particular instance. In the early years of courtroom DNA application, opponents challenged the results as unreliable, as happened in the famous O. J. Simpson trial, and juries could be convinced that the evidence was unreliable. Although DNA evidence can be, and often is, challenged on the grounds that the means for collecting and processing the samples were faulty, that is quite different from challenging the science on which the test is based. Just as other unimpeachable witnesses need to establish their qualities of reliability, so science must convince judges and juries that its methods are reliable, statistically sound, and productive of true conclusions. My concern is, in part, how that process of establishing credibility takes place.

There are many other instances in which science exerts authority in society's affairs. For example, think of the standards of weight and measurement. With the passage of time, standards for length, mass, time, and other quantities have become extremely precise and thoroughly grounded in sophisticated methodologies. The modern world is absolutely dependent on these standards, which have been set by a general agreement according to the advice of science. Or think of the innumerable tests, vaccines, and treatments employed in modern medicine, nearly all accepted without question as safe and effective: people by the millions take annual flu shots or consume cholesterol-lowering drugs.

Some scientifically validated procedures will generally be subject to vigorous challenge when the results could determine criminal guilt or innocence. The measurement of blood alcohol level in a driver suspected to be drunk provides an example. The breathalyzer test used by law enforcement officers in the field has an inherent uncertainty, which may call into question whether the defendant was in fact over the legal limit of blood alcohol content. For this reason, the defendant may ask for a more accurate test, based on a blood sample. This test, however, is also subject to errors of various kinds, such as procedural errors or improperly calibrated equipment. The challenge is not primarily to the veracity of the underlying science, but rather to the details of how the test is conducted. It is a matter of law that a particular level of blood alcohol, set by each state, is taken to represent legal intoxication. The appropriate level set into law is judged with reference to studies of performance impairments of human subjects as a function of blood alcohol content. The underlying idea behind the test and the standards set for it are based on scientific principles and experimental results. Society at large accepts them on the grounds of scientific authority.

In some instances a minority may view certain procedures and practices as grounded in faulty science (for example, fluoridation of public water systems, or an Environmental Protection Agency determination of the allowable level of a pollutant in a water supply). Nonetheless, while in specific cases there may be demurrals regarding particular widespread rules and practices, science exerts enormous influence on many important aspects of modern society. In so doing, it exercises a form of authority, even though it may go unnoticed or taken for granted.

## WHY A BOOK ABOUT THE AUTHORITY OF SCIENCE?

It is a cliché to say that science and technology have transformed life on this planet for a substantial proportion of its human inhabitants. The transformations, however, are not always for the better, whether one is considering those who live in the megacities that dominate life in the developed and developing nations, or those whose lives in previously remote places have been uprooted. Arguably, science and its sister, technology, have made life much better for most people, even as challenging problems have arisen. What does seem clear is that the world now cannot do without science and technology (Hughes 2004).

Every day, in communities all over the world, scientific and technological considerations are playing vital roles in business, medical, educational, and governmental programmatic and regulatory decisions. But how are those decisions made? What ideas, values, and motivations prompt them? To what extent are considerations drawn from science itself invoked in making each decision? Even as we acknowledge the indispensability of science to modern life, its place in society as a social force is continually questioned. Today, science seems to be pressed on several fronts to defend its place and to justify in economic, social, and moral terms its effects (as well as those of technology, to which it is symbiotically related). The struggle for societal acceptance and influence often pits science against other societal interests, including those that originate in business, religion, law, ethics, or a wide range of governmental policies. It can be seen as a continuous series of contests for authority, some of which science wins and some of which it loses. But what do the different outcomes hinge on? Is there something about how these conflicts work out that revels an underlying structure and set of values, or are socially adventitious factors at work?

The story of how science operates in society, how it interacts with other social forces and is evaluated, can be told many different ways. We cannot achieve a balanced understanding by thinking of science as just a special economic or social interest group, a domain of knowledge and practice with special privileging characteristics, or simply a resource for solving many of the world's problems. Each of these and similar conceptions of science reflects some truth about science in society, but fails by itself to capture the diverse nature of science's relationships to other societal sectors. Each thus fails to fully account for the unique place of science in modern life. By reflecting on science as a source of authority, we can achieve a more complex and balanced understanding of its roles in society. By authority I mean expert and moral authority, not coercive authority expressed in the form of law or power. In this way of looking at science, or any other societal interest group, authority is a measure of the capacity to instill belief; to engender not only understanding, but also assent; to move those affected toward changed attitudes; and to encourage actions.

Although there is a general recognition that science has authority to speak on some issues, most people are not clear on how that authority arises. My aim in this book is to clarify the sources of scientific authority, to identify its historical origins, and to show how that authority is continually challenged from various directions. I want to delineate not only the nature and origins of scientific authority, but also to consider its limits. Looking at science in society through the lens of authority requires that we reexamine the past in order to see how the rise of science in Western culture originated as a process of wresting authority away from other sectors. This process has advanced greatly, if not steadily, to the present day. For example, the contest between science and religion goes on today and is a major obstacle to full social acceptance of many theories and experimental findings widely accepted within science. In similar ways, the contributions of science to the foundations of many social policies and legal and ethical practices often give rise to conflicts with older traditions. For example, think of the influence of scientific findings in genetics and the behavioral sciences on our understanding of race, gender, the reliability of eyewitness memory, alcohol addiction, and many more such topics, and all their attendant implications for social policies and practices.

By looking at science's influence on society in terms of its expert authority, and in some instances its moral authority, the discussion inevitably turns to the autonomy of science. This is an important aspect of our subject because there is a strong coupling between the authority with which science can pronounce on many topics, and its intellectual independence and freedom. Yet autonomy can also lead to conflicts for science by opening up possibilities for research into socially unacceptable areas, or by generating demands for more resources than society is prepared to make available.

The question of if and how science can or should exert *moral* authority seems not to have been adequately addressed. Moral authority is distinct from expert authority; it is a measure of the capacity to speak convincingly about what *ought to* be, as opposed to what is. It is obvious that science and scientists attempt to exert moral authority all the time. One need only look at the editorial pages of prestigious journals such as *Science* or *Nature* to find dispensations of general advice, or strongly held positions on various controversial topics. The distinction between expert authority and moral authority, however, is often unclear. When a climate scientist talks about global warming, for example, does she simply report on the results of empirical studies, analyses, and model building, or does she argue, explicitly or implicitly, for actions that might be taken to ameliorate global warming's effects? What part of what she says can be considered an exercise of expert authority, and what part an exercise of moral authority? Similar questions arise when scientific organizations or individual scientists participate in public discussions of such problem areas as population control, nutrition, biomedical research involving stem cells, or the fight against diseases such as AIDS or tuberculosis. Regardless of the context in which it might arise, one can ask just how science comes to have moral authority, if indeed it does, and the extent to which that moral authority derives from the nature of science itself.

## WHAT THIS BOOK IS ABOUT

The expert, or epistemic, authority of science and the moral authority of science are distinct but strongly coupled ideas. To understand how science wields either expert or moral authority, we will have to pursue the following aims.

First, we need to clarify the ideas of expert authority and moral authority, particularly as they apply to science. What do we mean when we talk of authority generally? What are the limits of science's authority, and how do those limits arise? How does the authority of science differ from that exerted by other social entities, such as law, religion, or government? In what way is moral authority an extension of authority that arises from expertise? Just what are the grounds on which science can claim moral authority?

Second, we must outline the historical origins of science's authority. How did Western science come to possess authority? If one thinks in terms of a continuing competition for authority among various sectors of society, what sectors gave way to the growing influence of science? How was that shifting balance of authority expressed through the images of science that held sway during the development of science through the Enlightenment and to the present? How has the growth of science in the United States and its development as part of American culture contributed to its present position?

Third, we are to identify the bases of science's authority in contemporary society. Here we need to analyze how the enormous growth of science and technology, and their overwhelming effects on modern life, have influenced the ways in which science's authority is exercised, and—perhaps more important—how its authority and autonomy are constrained. Precisely because science and technology are so deeply woven into the fabric of modern life, they are more visibly subject to constraining forces than in earlier times.

Finally, we must trace the relationships of science to society using the concept of authority. Focusing on science's expert authority and moral authority in society affords a perspective on how science's societal roles open opportunities to exercise that authority. The enormous growth in science, and its correspondingly greater influence on societal affairs, has occasioned much more attention on what science produces and how it functions. There is no avoiding its presence in the courtroom, in the halls of government, or when matters of health and education are under consideration. At the same time, society is in a position to implement constraints on science through a wide range of actions that go beyond merely controlling levels of funding. How might such constraints change the nature of scientific inquiry itself, and thus alter the nature of the social contract between science and society?

Before we begin to tackle each of these aims, it is important to get straight some definitions and usages, and to establish the boundaries of our inquiry.

### TERMS AND BOUNDARIES

In what has been said so far, I have referred sometimes to science, and sometimes to science and technology, treating them as an interdependent pair. In the most simplistic way of looking at it, science is the study of the natural world using a characteristic mode of approach sometimes referred to as "the scientific method." It is widely appreciated that there is no single scientific method, but rather a family of methodological forms of inquiry that share a set of characteristics and have as their aim the production of new knowledge. That knowledge may consist of newly acquired and organized data, new theoretical insights, a new model of some aspect of nature, or a new method for conducting certain experiments. I will have more to say later about what counts as science for our purposes.

Technology, on the other hand, has as its aim the production of something of practical use: a new tool for science, a new way of carrying out complex business calculations, a new kind of glass for high-rise buildings, a new and more effective heart defibrillator. Technology flows from scientific discovery, so in some sense it is posterior to science. Because technology is the source of many of the tools that scientists use in their work, however, it is essential in acquiring certain forms of new knowledge. Indeed, much technology that finds its way into general use results from initial development of tools for use in basic scientific research. Thus, the results of science in fields such as physics, chemistry, and biology are closely coupled to technological applications.

Technology is an all-pervasive presence in modern society, and it is often the case that nonscientists confuse it with science itself. Most scientists, particularly those that form what we might call the "academic science" establishment (scientists working in academic research universities, research institutes, or government laboratories with a strong component of basic research), are generally discomfited by this confusion (Wolpert 2005). For the most part, it is technology that produces effects for both good and ill that are widely perceived throughout society. Technologies seen as beneficial help to justify investment in basic science. Conversely, when the technology is perceived as inimical to society's interests, public support for science funding may decline, or there may be calls for policies that restrict the conduct of research. Finally, technology is most often the product of commercial enterprises or large-scale government programs, such as the Radiation Laboratory during World War II. Thus, science and business are inextricably coupled, as are science and national policy.

To a large extent, it is through technology, in the form of material changes, that science affects society. Therefore, one cannot ignore technology when considering the place of science in society. But the justification for speaking with authority on a given topic most often rests with science, for reasons that we will be discussing, even though a given issue may rest on a point of technology. As such occasions arise in our discussions, the specifically technological nature of the issue will be highlighted. In the interests of a simpler language, I will generally refer simply to science rather than science and technology.

Science is by no means a monolithic enterprise. In the first place, various kinds of activities and entities count in some respects as science (Ziman 1984).

*Basic research* is curiosity-driven, more or less independent of any foreordained applications, and conducted with the intent of uncovering new understandings of nature. It may involve the design and performance of experiments, analysis of data collected, formation of hypotheses based on the observations, development of models consistent with the observations and data, and theoretical representations of those models. It also includes the dissemination of scientific findings through presentations at confer ences and invited lectures, but primarily through publication in scientific journals.

*Applied research* is conducted with the aim of applying scientific find- ings to the solution of a specific problem of interest outside the immedi- ate sphere of the work (for example, research to develop a new anticancer therapy). Clearly, applied research is intended to lead to new technolo- gies, broadly speaking. It is the sort of research conducted in industry, where the focus is on production of new knowledge that furthers the interests—usually rather short-term—of the company. Dissemination of results is an element of importance in applied research, just as in basic research, but considerations of intellectual property rights often limit pub- lic disclosure.

As opposed to the activities of research, *scientific* knowledge is embod- ied in a scientific literature that reports the results of research and the products of scientists' thoughts. This literature provides a record of already- published work, is dynamic in content, allows for new results and inter- pretations to continually replace older ones, and facilitates entirely new areas of investigation being reported on and opened for further work.

*Science education* is the teaching of science at all levels within the edu- cational system, including graduate education. The teaching may involve imparting information about the content of a particular area of science (e.g., astronomy or organic chemistry), but may also be about the nature of scientific activity more generally: methodology, how to perform experi- ments or use particular techniques, and so on.

*Science writing* describes the activities of scientists and the results of their experiments, and may be disseminated for a larger audience through media such as TV, books, newspapers, magazines, and the Internet. Sci- ence writing may attempt to convey the nature of scientific work or the sig- nificance of particular results. It is meant primarily for nonspecialists and the lay public.

A mere categorization such as the preceding, however, does not go far toward capturing the complexity of modern science. In addition to the tra- ditional fundamental scientific disciplines such as mathematics (which we count as science because of its central position as a lingua franca in so many scientific fields), physics, astronomy, chemistry, biology, and all the subfields of those large-scale divisions, there are more applied fields, such as geology, oceanography, plant science, dairy science, and materials sci- ence, devoted to work in specific areas. Then there are the various fields of engineering, such as civil engineering, aeronautical engineering, and electrical engineering, all of which draw heavily from the physical and natural sciences to form the core of the individual disciplines.

Alongside the natural sciences and engineering, there are the social sciences, which include disciplines such as economics, psychology, and sociology, along with interdisciplinary areas such as cognitive science, an umbrella term for a large grouping of disciplines that includes parts of linguistics, philosophy, neuroscience, computer science, and psychology. Medicine, which has its own distinct ethos and practices, increasingly has become "scientific," as evidenced by its heavy reliance on technology, and through the extraordinary impact of new findings in the basic biological sciences on the understanding of health and disease.

This is by no means a comprehensive cataloging of the scientific enterprise. Suffice it to say that the many fields of science and modes of intellectual pursuit vary greatly in terms of aims and methods, and most especially in their social structures. The individual areas of science are typically embedded within differing kinds of institutional structures, receive their support from differing sources, and are accountable to different societal entities. They vary in their traditions, standards of practice, reward systems, and means of internal communication. In addition, they may not be consistent in terms of what counts as acceptable scientific results. Partly because of such differences, the characteristics of those who practice these various fields of science also differ, in terms of personality type, career aspirations, and motivations.

In light of these considerations, is it possible to speak coherently of "science"? Is there anything in common between the director of a generously endowed medical research foundation; a professor of plant sciences in a college of agriculture; an M.D.-Ph.D. biomedical researcher in a large, research-oriented medical school; an environmental scientist in a department of civil engineering; a social scientist studying the factors contributing to the spread of AIDS in an urban environment; a scientist working in a corporate research laboratory; a program officer in molecular biology at the National Science Foundation; and a graduate student in a neurosciences program?

I will not attempt here what may in any case be impossible: to produce a kind of litmus test for what constitutes "science" or "scientist." Nevertheless, I think that we can draw some boundaries, albeit rather flexible or even in some cases indistinct, around a body of knowledge and current practice that represents a characteristic outlook toward the physical world, acceptable approaches to the study of nature, an avowed ethic of practice, commitment to critical evaluation of findings, and a commonality of social practices of communication, such as peer-reviewed journal publication. Those boundaries would include the traditional fundamental areas such as mathematics, physics, chemistry, the biological sciences, and the interdisciplinary and applied sciences that derive from these. They would also include much engineering research and the social sciences to the extent that they also embody the characteristics mentioned above.

For our present purposes, scientific practice does not include much applied research, such as process development or product testing; writing or speaking about science; advocacy for science or technology in legislative circles; teaching of science; medical practice; or administration of a science and technology program in a funding agency. Someone with bona fide scientific credentials might very well carry out any of these activities, but the activities themselves are not science in the sense that I would like to think of for our purposes here. A person engaged in any of these activities might speak with scientific authority, but it would necessarily arise from the person's professional credentials and not out of their activities. Ambiguous cases are bound to arise no matter how we distinguish between someone who can claim scientific authority and someone who cannot. Further, it is certainly the case that many who are not practicing science, but who write or legislate about it, or who administer scientific programs and projects, do have important influences on the authority of science (for example, through their influences on public perceptions of science).

From the perspective of some sociologists who study science, establishing the boundaries of what constitutes science—"boundary-work"—is a continuing process; the "space" of science is a matter of continual negotiation:

> "Science" is a cultural space: it has no essential or universal qualities. Rather, its characteristics are selectively and inconsistently attributed as boundaries between "scientific" space and other spaces are rhetorically constructed. The longstanding question, "What unique, essential, and universal features of science justify its authority in politics, law, media, advertising, and everyday reckonings of reality?" should be replaced, I suggest, by this more tractable question: "How do people sustain the epistemic authority of science as they seek to make their claims and practices credible (or useful) by distinguishing them from unworthy claims and practices of some nether region of *non-science*?" (Gieryn 1999, xii)

Many scientists would like to think that society's appreciation of science's "unique, essential, and universal features" are at the core of the field's special epistemic status. But there is no denying the practical realities that underlie Gieryn's suggested phrasing of the question regarding the source of science's authority. Nonscientists generally are not interested in finely drawn discussions of epistemology. They need to be convinced on less esoteric terms that science is different from other intellectual arenas, and that it can speak with authority on many matters of societal concern. For anyone speaking for science, that means explicitly distinguishing science from other forms of knowledge and practice—that is, establishing the "territory" of science—in terms that nonscientists can readily comprehend.

## WHAT LIES AHEAD

The book is organized into two major divisions. The first, comprising chapters 1 through 4, is concerned with foundational materials. In the first chapter I consider what authority generally means: what kinds of authority there are, how it is vested, and the limits that might apply to it. We will also examine the distinction between authority and moral authority before turning to the issue of authority as it applies to science. Does scientific authority inhere in individual scientists, in science as an institution, or in both? On what grounds can science claim to exert authority? What are its limits? Can science itself claim to exert moral authority, or does that flow from something else? How is authority related to autonomy? These questions are at the heart of our inquiry and will arise repeatedly in the chapters that follow.

Chapter 2 is a brief historical account of how scientific authority grew in Western society, beginning with the Ionian Greek thinkers and proceeding through the medieval Christian era. In terms of authority relationships, Galileo represents a kind of watershed. Those who followed him, such as Gassendi, Mersenne, Hobbes, and Descartes, helped to establish a rational authority for science and in the process responded to the challenge of radical skepticism. The Enlightenment period in England ushered in the beginnings of organized science in the form of the Royal Society of London. Science began to have a visible social presence that grew over time and in other nations. I conclude the chapter by examining the career of Louis Pasteur, the charismatic nineteenth-century scientist who immersed himself thoroughly in the affairs of society.

Chapter 3 deals with the rise of American science. The American culture impressed a unique character on its science: a pervasive tension between autonomy and service to society, and between religious covenants and scientific assertions about the natural world. In the early twentieth century, the advance of science was coupled to ideas of national progress. World War II was of particular importance in enlarging science's roles in national life. Science was promoted as a mainstay of national defense during the cold war era, and at the same time touted by some of its advocates as a kind of ideal social community. But the tensions between scientific autonomy and conflicting societal expectations remain active to the present day.

Chapter 4 is an analysis of science's present position in society, the claims made by science to objectivity, and its capacity to attain truths about nature. To properly understand how science operates in society we must understand its internal social structure. In the wake of Thomas Kuhn's groundbreaking 1962 work, *The Structure of Scientific Revolutions*, some social scientists adopted controversial new views of the nature of scientific activity. With time, more nuanced models, with deeper appreciations of the roles of testimony and trust, have evolved. The ideals of impersonality, objectivity, consensus, and disinterestedness form the major ingredients of science's public face. The chapter concludes with a further look at the concept of moral authority.

The second major section of the book contains detailed examinations of the roles of authority in science's interactions with various major sectors of society. In chapter 5 I examine the contrasting meanings of truth in law and in science, as well as the means for establishing it. A critical issue for the authority of science in law is the admissibility of scientific evidence. The long history of fingerprint identification testimony provides insights into the establishment of authority in the courtroom. I contrast that with the establishment of authority for more modern methods, such as DNA fingerprinting. The landmark Supreme Court *Daubert* case, and others that followed from it, have profoundly affected science's place in the courtroom. The rules of evidence in their present form invest judges with a powerful gatekeeper function in deciding which evidence, purporting to rest on valid science, is to be admitted into testimony. It does not appear that present practices are up to the challenges of dealing with rapidly changing science, however, nor do they provide plaintiffs in many cases with fair and equitable access to relevant studies.

Chapter 6 deals with the conflicts between science and organized religion, which each derive their authorities in different ways. Science exercises epistemic authority grounded largely in rational-legal foundations, and religion exercises epistemic and moral authority rooted mainly in tradition. In American society the conflicts between science and religion have been pervasive and continue unabated, despite the growing influence of science in society generally. I explore the nature of these conflicts in three topical areas: evolution, human reproduction, and stem cell research. In all these cases the conflicts are played out in the public media, courtrooms, regulatory agencies, and in federal, state, and local legislative and executive bodies.

Chapter 7 is concerned with science's authority relationships with government. Since the end of World War II, the protracted cold war period and the growing power of science have forced new examinations of issues such as the rationales for funding scientific research, the uses of science in governmental affairs, the effects of scientific influences on governmental policies, and governmental constraints on scientific research. The influence of scientific authority on governmental policies, as well as powerful resistance to its exercise, are exemplified in recent and current topics: the politics of stem cell research, stratospheric ozone depletion, and global warming. Science's authority and autonomy have been compromised on many occasions by policies and practices of the executive branch and powerful congressional committees.

If science is to play a significant role in the affairs of society, it must connect effectively with the public. The nature and extent of those connections are examined in chapter 8. In the 1970s, concerns that general harm might result from research involving then-new recombinant DNA brought together scientists and members of the public and local governments in Cambridge, Massachusetts, and elsewhere. The episode illustrates how science can effectively maintain an epistemic authority in advancing its aims when scientists attend to the public's concerns and respond appropriately. More generally, the science-public nexus depends on the extent and quality of the public's sources of scientific knowledge, and is conditioned by public views of moral and ethical issues surrounding science, including research misconduct, misuses of scientific knowledge, and conflicts of interest and commitment on the part of researchers. Because science competes with other social entities for attention and support, low scientific literacy on the part of the lay public is a cause for concern. I argue that a broader public understanding of the means by which scientific knowledge is made, of science as a social institution, and of how scientific authority contrasts with other sources of authority, rather than knowledge of specific theories and bodies of descriptive knowledge, is key to science's more successful interactions with the public.

In chapter 9 I have attempted to synthesize the various insights and conclusions derived from the earlier chapters into a coherent picture of science as a source of authority. This picture paints science as a powerful presence in society, but one with limited expert and cultural authority in contributing to public culture. Science is generally valued highly for its instrumental contributions to human welfare. Indeed, it is the implementations of these that largely form the basis for science's cultural presence. But science has not established an image that serves well when it comes into conflict with other claimants to authority. It frequently fails to convince nonscientists of the truth of well-established scientific claims. It also frequently falls short in its attempts to exercise moral authority on contested issues. The difficulty lies partly with science's organization along the lines of a self-contained "republic," in which controlling standards of performance and evaluation place a low priority on the participation of individual scientists in societal affairs. Further, scientific rationality has limited standing in society at large as a process for resolving the many issues that vex modern life. These and other considerations call for a reappraisal of the tools and strategies employed by science and scientists in interfacing with other societal sectors, especially the general public.

# I
## FOUNDATIONS

# ONE

Before anything significant can be said about the authority or moral authority of science, we need to be clear on what these words denote. There are several types of authority. Joseph Raz, in his introduction to the 1990 volume *Authority*, writes: "An authority on medieval coins, or on Chinese eleventh-century porcelain, or on quantum mechanics, or on aerodynamic properties of some new materials: these are all people who are expert in their fields, i.e., who are good at stating how things are. Their judgement is a particularly reliable guide as to how things are, independently of that judgement" (2). He adds: "To have authority . . . is to be an expert who can vouch for the reliability of particular information. This authority of the expert can be called theoretical authority, for it is an authority about what to believe."

We all defer to the authority of experts in our everyday lives. We count on physicians to correctly diagnose our ailments, on the meteorologist to give us a reasonable forecast of tomorrow's weather, our financial advisers to steer us clear of imprudent investments. In doing so we do not generally ask for a detailed description of their line of reasoning, though we may be reassured by a modicum of explanation. It is just good common sense to defer to the authority of experts in many areas of life where we lack the background and insights to make decisions on our own. We do, however, require a sense of surety that the expert is really knowledgeable and trustworthy. When we are confronted with having to choose between opposing authoritative views of a matter, we generally end up choosing a particular one on the basis of trust. In looking for the origins of science's authority, therefore, we will be searching for what it is about science that engenders trust.

In his book *Science in a Social World*, Alvin Goldman offers a definition of an authority in characteristic philosopher lingo: "Person A is an authority in subject S if and only if A knows more propositions in S, or has a higher degree of knowledge of propositions in S, than almost anybody else" (1999, 268). Authority is defined here as depending on knowledge held, in terms of what is known as well as its significance. Goldman is quite willing to enlarge this definition to include the intellectual skills needed to answer questions on a given topic, even if the answers have not already been found. In a case of conflicting advice from sources that claim expertise, and in the absence of any other grounds for making a distinction, we would be inclined to trust the source that we deem to have the greatest expertise with respect to the matter at hand (Goldman 2006, 14–38).

There is often no foolproof way to establish expertise, however, or to be sure that we have not been led by various social and cultural influences to accept bogus claims (Walton 1997, chap. 1). Suppose we have a difficult issue facing us, and there are many ramifications attending either of two potential solutions. As prudent people should, we seek expert advice on how to proceed. But how do we know when we have found a reliable expert? We may rely on reputation, credentials, demeanor, and other characteristics, but because we are, by definition, not expert in the matter at hand, we are hard put to feel that we have rational grounds for making the right choice. If we consult another expert for advice on the first expert, we have simply moved the question back one level: how can we know that this second expert is a reliable authority? There is no simple answer to this dilemma; in practice we acquire beliefs and knowledge through the testimony of others and through surrender of some of our epistemic individualism (Hardwig 2006, chap. 11).

While authority generally may be essential to the life of society, it is not always seen in a favorable light. In whatever form it appears, there is in some sense a coercive or suppressive quality about it; it militates against actions and thoughts that differ from those promoted by the authority itself. For example, with respect to truth:

> Our daily life is constantly troubled by vexing questions, ideological, ethical, political, esthetic, and factual, to which we cannot remain

indifferent, to which we must give some sort of answer, and which involve such obscurities that an answer in terms of objective determination is very hard to reach. But most of the time these questions admit of cheap, easy, pacifying, and heartening answers if we make it a set rule to repeat what authority has said. The lovers of truth easily come to suspect that the whole system of authority is a pragmatic device, designed to spare weak souls the hardship of finding truth and abiding by it. (Simon 1962, 16)

Put another way, it is one thing to establish authority; it is another to have it received gladly by those at whom it is directed. This can apply, though perhaps in slightly different senses, to the authority of the scientist speaking on a contentious issue, such as stem cell research, as well as to that of a police officer handing out a parking ticket. That comparison brings us to a consideration of the various forms of authority.

## KINDS OF AUTHORITY

We have so far been considering authority grounded in expert knowledge, whether of facts or methods. Authority of this kind is referred to variously as "epistemic," "intellectual," "expert," or "cognitive." The sociologist Max Weber, however, viewed the concept of authority in broader terms. He advocated a strongly methodological and empirical approach to the study of the sociological and institutional foundations of society, and can be fairly counted as one of the founders of social science. The idea of authority plays a dominant role in his thought and writings. In the late nineteenth century, he presented a taxonomy of authority in *The Theory of Social and Economic Organization*, in which he saw authority as having three sources (Weber 1915, chap. 3).

In Weber's scheme, *rational-legal* authority derives from the body of generalized rules that apply to an organization, stipulating its powers and the limits of its operations. The organization might be a governmental unit, a corporate entity, a scientific society, or other social entity. Authority is implicit in the order of the organization and does not derive from the particular characteristics of whomever might occupy a given position. Rational-legal authority usually involves a bureaucratic structure, which comprises delineated spheres of authority for those occupying various positions. Authority is conferred on individuals occupying offices or defined roles within the social structure: the teacher, police officer, or chair of an appointed scientific committee.

*Traditional* authority is possessed by the pope, a king, or the father of a family. It rests on long-hallowed tradition. Weber says that authority "will be called 'traditional' if legitimacy is claimed for it and believed in on the basis of the sanctity of the order and the attendant powers of control as they have been handed down in the past, 'have always existed.' The person or persons exercising authority are designated according to traditionally transmitted rules. . . . Obedience is not owed to enacted rules, but to the person who occupies a position of authority by tradition or who has been chosen for such a position on a traditional basis" (1947, 341).

Weber characterizes *charismatic* authority this way: "The term 'charisma' will be applied to a certain quality of an individual personality by virtue of which he is set apart from ordinary men and treated as endowed with supernatural, superhuman or at least specifically exceptional powers or qualities. These are such as are not accessible to the ordinary person, but are regarded as of divine origin or as exemplary, and on the basis of them the individual concerned is treated as a leader" (358). Action heroes such as Rob Roy, the Scots freebooter made famous in Sir Walter Scott's eponymously titled novel, or George Washington, chief of the Continental forces in the American Revolution, are exemplars of charismatic authority. Extraordinary intellectual accomplishments or leadership abilities can serve to endow individual scientists with charismatic authority, both within the science community and in society at large. Examples include Isaac Newton, Louis Pasteur, Albert Einstein, and Linus Pauling.

Apart from distinguishing authority in terms of its social origins, as identified by Weber, there is another dimension, having to do with the *nature* of the authority exercised. For our purposes, it will be most helpful to think in terms of three forms.

*Coercive* authority is usually exercised de jure, on the basis of laws, rules, tradition, or dictatorial powers. The police officer handing out a parking ticket, the judge deciding on a divorce settlement, and the Catholic Church's prohibition of contraceptive use by its members exemplify the exercise of coercive authority. Coercive authority does not apply to science except under socially pathological circumstances. Forced acceptance of scientific doctrines can occur in totalitarian regimes, as with Lysenkoism in the Soviet Union during Stalin's regime. The agronomist Trofim Lysenko propounded new agricultural practices based on scientifically unsupported theories of heredity. For political reasons the Soviet mass media portrayed

Lysenko as a hero and savior of Soviet agriculture. Rival views were rejected because they were seen as "bourgeois" or "fascist." Scientists who publicly opposed Lysenko's views were systematically liquidated. There was a wholesale purge of geneticists throughout the Soviet Union. I will not be concerned with such cases; coercive authority will come up in our discussions only insofar as scientific authority may stand in opposition to it.

*Expert* authority is the capacity to convince others of how the world is. It is what first comes to mind when we think of scientific authority (Goldman 1999, chap. 8; Kitcher 1993, chap. 8; Longino 2002, chap. 3). A person recognized within the scientific community as an outstanding entomologist, with long experience in the study of honeybees, exercises expert authority when offering an explanation of a recent precipitous decline in honeybee populations. Her long training and scientific accomplishments induce not only other scientists, but also society at large, to accept her claims about how things are within this narrow field of knowledge. An essential element here is trust both in the knowledge and integrity of the individual scientist, and in science as a social institution. Other terms, notably *epistemic* authority, *cognitive* authority, or *intellectual* authority, rest on largely the same set of implicit assumptions about the trustworthiness of scientific knowledge. I will have more to say later about *cultural* authority, a term favored by some who are concerned with sociological accounts of science's place in society (Gieryn 1999). It is meant to capture the fact that science's credibility derives from myriad relational ties that render it visible at many levels of society.

*Moral* authority is the capacity to convince others of how the world *should be.* You may not find this simple definition very satisfying, and I will have to work hard to convince you of its merits. The term moral authority can mean many different things, depending on the context of its use. Enter "moral authority" (within quotation marks) into a Google search and you can expect somewhere in the vicinity of 900,000 hits. The simple definition I propose encompasses the capacity to motivate others to work toward achieving goals that will change the world in some desired way. Addressing the International Conference on Climate Change in Bali in 2007, former U.S. vice president Al Gore exercised moral authority in urging governments to take steps to curb carbon dioxide emissions. Similarly, the president of the United States exercises moral authority in urging congressional leaders to act on particular pieces of legislation. These two examples illustrate the varied relationship between the *source* of authority on the one hand, and the *nature* of the authority exercised on the other.

Gore's role as a leader in advocating for action to mitigate climate change, recognized by the award of the 2007 Nobel Peace Prize just prior to the Bali conference, conferred the charismatic authority that gave weight to his advocacy. (His arguments rest, of course, on a vast amount of expert authority in the scientific community on the subject of climate change.) By contrast, the source of the president's moral authority in his dealings with Congress is his rational-legal authority as the elected president.

## INSTITUTIONAL VERSUS INDIVIDUAL AUTHORITY

I have already suggested that particular individual scientists can possess charismatic authority. But can we think of the authority of science generally as residing in individual scientists? Or is it a property of science as an organized activity, a social enterprise that exerts social effects distinct from that of individuals? One could argue that an individual scientist can never exercise authority without drawing on multiple ties to science as an institution. The influential sociologist of science Robert Merton viewed the authority of science as being vested not so much in individuals as in science as a coherent social entity, possessing characteristic norms and institutional checks and balances (1949, chap. 17). Recognition of the expert authority of the entomologist of our earlier example depends on her bona fide membership in a relevant scientific community. Her expertise was developed in the culture of a large, highly organized scientific enterprise. Her scientific work receives validation through her career advancements, acceptance by the scientific community of peer-reviewed publications, and other evidences of success, such as awards. Finally, whatever she may have to say about bees will inevitably draw on the work of many other scientists.

The authority of science is a form of rational-legal authority in Weber's taxonomy: it is grounded in the workings of an established social institution that has its own organizational structure and standards, credentialing mechanisms, and procedures for assigning roles. To the extent that this is an appropriate way of characterizing matters, to understand the authority of science we need to examine the roles of scientific institutions in society as a whole (Bender 1984, 84–106).

Even though the institutional structure of science is essential in creating and sustaining authority, its exercise by individuals within that structure is frequently its most salient aspect. The extent to which an individual scientist can speak or write with authority, however, depends on the context. The scientist must be widely accepted as having not merely expertise, but also wisdom and judgment. Furthermore, the social sphere within which an individual scientist's authority is effective can vary widely. Famous persons such as Pasteur, Einstein, or Pauling were able to cast their charismatic influence across many sectors of society. But scientists or science spokespersons are often influential in more local settings. For example, a faculty member of a university may have a favorable reputation within the local community and be accorded respectful attention on a variety of issues. To exercise authority on a larger stage, exceptional measures of achievement are needed, such as major scientific prizes or widespread favorable notice of a scientific result. For example, this snippet from a press release illustrates the star power of the Nobel Prize: "Thirty-two Nobel Laureates and industry leaders wrote to President George W. Bush today to urge increasing funding for physical sciences, environmental sciences, mathematics, computer science and engineering" (APS 2003).

## JUSTIFICATIONS FOR SCIENTIFIC AUTHORITY

Authority may be represented by inanimate things, such as a crown and scepter, a book of laws, or a physics textbook, but it has meaning only when exercised. The prime minister might exercise her rational-legal authority in making appointments of cabinet members; the church may exercise its traditional religious authority in mandating particular beliefs, or in urging a particular choice of vote on a contested social issue; a scientist might exercise expert authority in declaring that the defendant was not the person who deposited a sperm sample at the scene of a rape crime, or in declaring a particular substance effective in treating a disease.

In some sense, authority is a capacity to wield power: to make things happen or to establish or change beliefs. Except for coercive authority, this capacity derives from de jure or normative strength, rather than from the naked power of money or physical force. Weber insisted that each type of authority in his taxonomy requires legitimacy, whether in terms of accepted tradition; a broad concurrence with, or submission to, the rules by which society is governed; or by virtue of popular acceptance. To comprehend the authority of science we will need to understand the processes by which it gains legitimacy, against what other sectors of society it contests for authority, and the circumstances in which authority is exercised.

Science's claims of a distinctive expert, or epistemic, authority must rest on a set of special properties that distinguish it from other kinds of activity and other forms of social organization. Views of the nature of science, and thus the justifications offered for its claims of authority, have varied over time. The idea of science in the eighteenth century was vastly different than it is today. Yet, as it has developed, particularly in Western culture, certain self-assertions about its nature have figured persistently in society's image of science and scientist.

Science claims to practice a *rational, objective methodology* for the study of nature and for organizing our understanding of it. The consistent application of methods that have led over time to reliable, reproducible, and intelligible facts, hypotheses, and theories has lent science considerable cachet as a source of truth. To the extent that society accepts science's outputs— organized observations and data, methods, hypotheses and theories—as truthful or leading toward truth, science can claim epistemic authority. The validation of these outputs within science itself, however, is indispensable if they are to convey authority. In his classic 1962 paper, "The Republic of Science," Michael Polanyi sketches how scientific authority arises:

> Let it also be quite clear that what we have described as the functions of scientific authority go far beyond a mere confirmation of facts asserted by science. For one thing, there are no mere facts in science. A scientific fact is one that has been accepted as such by scientific opinion, both on the grounds of the evidence in favor of it, and because it appears sufficiently plausible in view of the current scientific conception of the nature of things. Besides, science is not a mere collection of facts, but a system of facts based on their scientific interpretation. It is this system that is endorsed by a scientific authority. And within this system this authority endorses a particular distribution of scientific interest intrinsic to the system; a distribution of interest established by the delicate value-judgements exercised by scientific opinion in sifting and rewarding current contributions to science. Science *is what it is*, in virtue of the way in which scientific authority constantly eliminates, or else recognizes at various levels of merit, contributions offered to science. In accepting the authority of science, we accept the totality of all these value-judgements. (68)

We will return in chapter 4 to the important question of how "scientific opinion," or consensus, is reached within science. For now I simply note that science is, among other things, a social organization composed of people with differing backgrounds, situated in the larger society in many different ways, and with various ways of relating to others outside science. Because scientists and those who purport to speak for science interface with the world in disparate ways, the "outputs" of science are not mainly the data, methods, hypotheses, and theories with which science works internally. For the most part, these outputs are packaged products of science's internal workings framed in terms of characteristic ways of viewing the world and analyzing and solving problems. The credibility of the statements of individual scientists, as well as what may be perceived as the position of "science" on a given topic, are conditioned by individual reputations, and by assessments of bias within the science establishment.

Some claim that science is inherently *progressive* in its accumulation of knowledge and in the techniques it employs for investigating and manipulating nature. There is a *directionality* to science; it advances—albeit by sometimes tortured paths—toward the ideal of an objective, literal understanding of the natural world. This advancement in understanding and practice has contributed culturally to a more profound understanding of our place in the universe, and practically to make more comfortable, healthier, and richer lives for those living in societies where scientific understanding is applied. During the past half century this ameliorant view of science has been under attack from many directions. Thomas Kuhn's *The Structure of Scientific Revolutions* (1962) is credited with catalyzing new thought about the ways in which science progresses and the kinds of claims that it can make. We will explore critiques of the privileged status of scientific testimony and its social roles in later chapters.

Some also claim that those who practice science are for the most part *disinterested* in their search for truth and largely free from parochial interests and biases. That is to say, the ideal in science is objectivity, which Helen Longino characterizes as "the willingness to let our beliefs be determined by 'the facts' or by some impartial and nonarbitrary criteria rather than by our wishes as to how things ought to be" (1990, 62). This characterization can be most convincingly advanced as the ideal of those who pursue fundamental, curiosity-driven research. Arguably it also applies to industrial research laboratories and various other situations in which scientific study is motivated by a non-fundamental curiosity. Scientists whose work is directed toward specific ends, including new products of commercial value, are expected to observe such ethical norms of science as use of all applicable tools for ensuring accuracy and precision, full reporting of results, and acknowledgment of prior scientific literature and the contributions of coworkers. In these senses, the term "disinterested" applies to all efforts at scientific discovery.

Some see the science establishment as stressing the image of disinterestedness as a means of masking the real performative elements that underlie its authority (Hilgartner 2000). Maintaining the aura of impersonality and objectivity is essential to science's cultural authority. Additionally, consensus within science is an important element in sustaining the credibility of scientific claims made in the public arena. We will see examples of the stresses occasioned by the impulse toward consensus as we examine the interactions of science with other societal sectors.

### MORAL AUTHORITY

We have until now been using the term "authority" in differing ways, sometimes identifying the distinctions by means of modifiers. With such wide and variant usage of the term in diverse discourse venues, a certain looseness in terminology is inevitable. It is important, however, to settle on usage within this book. Statements of individual scientists or groups of scientists, or a consensual expression of the science community more generally, dealing with properties of and theories about the natural world are all forms of scientific *epistemic* or *expert* authority. This is what I mean by use of the term "authority" in reference to science, regardless of whether I preface it with the word "epistemic" or "expert." I want to distinguish the idea of authority as I have just described it from *moral* authority. As an example of scientific epistemic authority, an expert on spider venoms might describe conclusions reached from research regarding toxins in the bite of the *Epeira sclopetaria* species of spider. She might describe the identification of various molecular components of the venom, determinations of their chemical structures, and tests of their toxicities. Her expertise induces us to accept her claims about how things actually are within this particular circumscribed field of knowledge. By contrast, a political figure, religious organization, or charismatic leader may argue for what *ought* to be, based on a claim to moral authority. The president of the United States exercises both rational-legal authority and moral authority, sometimes concurrently. For example, he has the legal authority to veto legislation; at the same time, he can take advantage of the bully pulpit of the presidency to present his reasons for a veto and to urge the Congress to pass a different bill.

Can science exert moral authority in this sense? In the history of Western science, it has claimed to do on many occasions. One has only to read the editorials in preeminent journals such as *Science* or *Nature* to see that science continually attempts to do so today. But what is the justification for science's doing so? Science is concerned with determining facts about the physical world and formulating causal explanations of the facts in the form of theories. But to exercise moral authority is to argue for what *ought* to be—that is, to assert that one state of affairs is to be preferred over another. Although it may not always be the case, such arguments typically amount to moral claims. Moral philosophers are divided on the question of whether moral authority can arise from facts alone (Gewirth 1978; Mackie 1977). At issue is whether it is possible to construe the difference between right and wrong simply from knowledge of the natural world, which is of course precisely the kind of knowledge that science possesses (Patzig 2002).

There are two questions here. The first is whether any distinction between moral right and wrong can have an objective, universal validity, independent of any religious or social traditions, as opposed to being historically contingent and context-related. If so, it should be possible to make moral judgments using rational methods on which everyone can agree. Even if one grants that a moral judgment can rest on rationally derived general principles, there remains the question of what it takes to make a rational moral judgment. For science the primary criterion is clearly expert authority. But beyond this, in the course of arriving at a statement that conveys moral authority, at what juncture does science per se—having delivered the facts, as it were—cease to play a role? Are facts alone sufficient for establishing moral authority? Most philosophers would probably argue that moral judgments cannot be made on this basis—that they are specific to particular events, human perceptions, social norms, and psychological contexts. Is there something about nature itself or our relation to it that somehow conveys moral authority? Or is there a sense in which scientific knowledge, when it assumes the form of "practical knowledge," is vested with moral content (Annas 2006, chap. 9)?

It is useful to distinguish between the moral authority of science as an institution and that of particular scientists. The moral authority of a scientist will rest in part on the same virtues we look for in evaluating the moral authority of any individual: honesty, cooperativeness, concern for others, discipline, and so on. Beyond this, however, does the individual posses a moral authority by virtue of being a scientist? And is there a moral authority for science as an institution in society? The answers may lie in whether science as an institution is seen to possess virtues of the sort that we ascribe to individuals.

Science can make normative claims regarding its contributions to culture and human welfare, including the following:

- The products of scientific activity can be beneficial for individuals and society generally. While counterexamples can be invoked against this claim, scientists can point to the enormous improvements in the health and standards of living of much of the world's population that have resulted from science.
- Science enhances human existence by answering to our innate sense of wonder and longing to understand the physical world.
- Science is a self-disciplining and self-correcting community. Abuses such as falsification of data, improper attribution, and other ethical improprieties certainly exist, but they are detected and corrected for important scientific results through collective action within the scientific community. Science can be seen as special in this respect, not because scientists are more moral or selfless than the run of humanity, but because of the way that science operates in terms of methodology and social structure.
- The scientific enterprise has a social stability that flows from the characteristic rationality of scientific practice and an emphasis on communication as a tool in achieving intersubjective agreements. Science has been more successful than most social enterprises in overcoming cultural differences among its practitioners. It offers a model of how to organize human activity in advancing social goals of a particular kind. Admittedly, however, this estimable trait applies to practices of narrow scope in comparison with the broad range of societal concerns.

Gerald Holton, in his Lewis Branscomb Lecture of 2000, identified what he saw as five components that make up the moral authority of science:

- *The imperative of excellence.* This component is integral with the claims of science to produce advances that are universally testable and acceptable.
- *Internal accountability.* This amounts to obeying communally held ethical imperatives in the conduct of research.
- *External accountability.* Science must be accountable to the larger society by explaining what it is doing and what it means. This also includes a commitment to science education.

- *Identity preservation.* Science must work actively to maintain its unique identity against the incursions of other claimants for authority. This entails "clarity within the scientific community what science is, what it is not, and where the limits are."
- *Communitarian obligation.* This involves "coupling science and technology to the wider national interest, doing basic research for the public good." Implicit in this element is surrender of a certain amount of autonomy, in that the choices of research topic are driven by something other than pure curiosity and evaluation of scientific merit.

Holton's five elements overlap to some extent with those I have already delineated. They reflect largely similar notions of what gives rise to moral authority.

It would be easy to conclude from much of the current literature on science's roles in society that it can be found wanting in one way or another with respect to any of these claims for authority or moral authority. But in fact many people, particularly many scientists, do accept some or all of these elements wholeheartedly, even as they acknowledge that science frequently falls short of its ideals. Yes, say the defenders, scientists are human like everyone else, so they make mistakes and commit ethical failures. The authority, and particularly the moral authority, of an individual scientist may be corrupted. But, so the defense goes, science is self-correcting; bad science and unethical behavior are rooted out. The authority of science as a social force is not thereby irredeemably compromised. Furthermore, science is too often confused with technology. Scientists generally have limited control of what others may do with the results of basic, disinterested research.

To put some flesh on the bare bones of these ideas, let us consider a specific example of present social importance and salience that involves the exercise of both authority and moral authority. The illustration is relatively unproblematic in terms of the use of terminology.

### REPRODUCTIVE TECHNOLOGY

Infertility is an all-too-common problem. Beginning in the 1970s, a variety of procedures and treatments, referred to as assisted reproductive technology (ART) procedures, have been developed to help women get pregnant. ART procedures have enabled the births of more than one million babies worldwide. In the United States, where one in six couples are said to have fertility problems, more than one hundred thousand ART procedures are carried out annually, resulting in approximately sixty thousand births. The fertility industry has grown quite large: consumers spend more than four billion dollars annually in attempts to overcome fertility problems. ART methods include:

- *Artificial insemination*, in which male sperm, which may have been subjected to certain "washing" procedures, are inserted into the woman's uterine cavity through a long, thin catheter.
- *In vitro fertilization* (IVF), in which fertility drugs are administered to the woman so that several eggs can be harvested from the ovary using a probe inserted into the vagina. The eggs and sperm are then combined in a petri dish. Between forty-eight and seventy-two hours later, the eggs are usually fertilized. These fertilized eggs, or embryos, which represent the first stage toward development of the fetus, are then reimplanted into the woman's uterus. This procedure leaves open the potential for multiple births.
- *Blastocyst transfer*, a variant of IVF in which the early-stage embryos are left longer in the culture to grow. This allows for greater selection of only the best ones, based on their growth and development to that stage. Because fewer of the best blastocysts are reimplanted, the risk of multiple births is reduced. Other specialized techniques, some of them variants of IVF, have also been developed, and new ones continue to appear.

The thread of authority seems easy enough to trace. Biomedical researchers, working first with animal studies, and then with humans in trial experiments, established the potential for the various methods. These researchers, employing several different technologies in the form of research tools, were engaged in basic science. The various procedures, such as IVF or blastocyst transfer, became technologies, with accepted standards of practice, on the authority of the researchers who carried out the first experiments and established their efficacy. Authority in this example has to do with such questions as what is occurring in the procedure, what procedural methods work best (e.g., washing of cells, thinning of the membrane of the fertilized egg before implanting, and so forth), the claims regarding the efficacy of a given procedure, the viability of resulting fetuses, and so on. Those deciding to avail themselves of these means of fertility enhancement do so because they accept the expert authority of science as to the safety and efficacy of the procedures, based on the results of scientific research.

But beyond the biomedical details of the procedures themselves, there lurk many ethical and religious issues. Some of these have to do with social and legal matters involving, for example, the use of donor sperm and donor eggs, or surrogate motherhood, in which a woman serves as the gestational carrier of an embryo that was formed from the egg of another woman. Another group of issues that I want to call attention to arises directly from the nature of the procedures themselves. One of the most important is an increased likelihood of multiple-infant pregnancies resulting from ART procedures. Typically, in IVF procedures, multiple embryos are reimplanted in the woman's uterus because the probability of any one of them going on to form a viable fetus is low. In other cases, fertility drugs may cause the woman to release multiple ova in an ovulation cycle. These procedures thus give rise to the potential for multiple-infant pregnancies. In November 1997, an Iowa couple, Kenny and Bobbi McCaughey, became the parents of the world's first surviving septuplets, courtesy of ART. The septuplets were conceived as the result of fertility drugs. When the couple discovered that Bobbi was carrying seven fetuses, they declined to reduce the number of fetuses through a selective procedure. As the result of being born prematurely, several of the septuplets have medical problems; for example, two of the children have cerebral palsy.

As is evident from the McCaughey example, ART procedures are not an unmitigated good. They are responsible for a very large increase in the occurrence of multiple-infant pregnancies, which in turn are associated with complications for both mothers and babies: miscarriage, premature birth, birth defects, higher rates of infant death, and deficits in both physical and mental development in childhood.

The employment of any of the ART procedures is thus freighted with ethical and moral issues. Any individual or couple using one of these procedures must face the prospect of having to make difficult decisions regarding how to respond to multiple-infant pregnancy. Should those offering ART services be subject to formal ethical guidelines with respect to informing and counseling patients, and even possibly declining to afford services to some who seek them? Who has the moral authority to speak on these matters?

The biomedical researchers who developed the various ART methods, and those who have gained long experience in practicing them, can speak with authority on what might occur in a given instance, and the potential consequences of electing a given course of action. But does their expertise in any way equip them with the moral authority to advise on the kinds of ethical and moral issues that we have identified? To locate the source of their moral authority we might look into their motivations in developing ART methods in the first place. Was it to provide assistance to those previously unable to have children? If they were prompted in their work by a concern for the needs of others, they might then as individual scientists have gained a measure of moral authority beyond that held by any other person of goodwill. This is not, however, a question with an easy answer. The connection, if there is one, between the scientist's authority as an expert and his standing as a source of moral authority is one that we will confront repeatedly as we look at specific arenas in which science interfaces with the larger society.

## AUTONOMY

The legitimacy of any form of authority, but especially moral authority, rests on the idea of autonomy. Before we can trust a person as a source of authority we must be assured that she is independent of external influences that might affect her judgment or testimony. It is precisely the absence of such assurance that often suffuses the electorate with deep skepticism about the utterances of politicians, who are beholden to various interest groups for campaign funding, or at the very least must bargain with other politicians to arrive at compromise solutions to a great many legislative issues. It is in the interests of autonomy that the justices of many higher courts, particularly the United States Supreme Court, are appointed for life and subject to impeachment only under the most stringent terms.

Autonomy in science operates at several different levels. First of all, it means the independence of individual scientists to pursue research of their own choosing, and to follow it wherever it may lead. Because of the social structure of science, such a freedom for individuals does not lead to chaos. The efforts of individual scientists are guided by norms within their fields, by their dependence on the work of others in advancing their own scientific agendas, by collaborative efforts among scientists, and by their ability to gain funding for their efforts, among other factors. Against this model of freely practicing scientists, control of the activities of a group by a single authority would lead to a loss of the creativity that the individual offers. Michael Polanyi was a strong advocate of autonomy for individual scientists. Writing in 1945 about the insidious effects of state control of science in the Soviet Union, which provided fertile ground for propagation of Lysenko's doctrine of heredity, he had this to say about restoring the proper functions of science:

> One thing only is necessary—but that is truly indispensable. It is only necessary to restore the independence of scientific opinion. To restore fully its powers to maintain scientific standards in respect of all their proper functions, in the selections of papers for publication, in the selection of candidates for scientific posts, in the granting of scientific distinctions, and in the award of special research subsidies. To restore to scientific opinion the power to control by its influence the publications of textbooks and popularizations of science as well as the teaching of science in universities and schools. To restore to it above all the power to protect that most precious foothold of originality, that landing ground of all new ideas, the position of the independent scientist—who must again become sole master of his own research work. (1945a, 149)

In the early history of Western science, when it was practiced by a relatively few individuals, and mostly by those with private means, autonomy was not a dominant issue. Today, however, the resources required to perform competitive scientific research are generally beyond the reach of individuals. Science is practiced, written about, and talked about almost entirely by people who are employed by others. An institutional setting for research, such as a corporate or governmental laboratory, or the facilities of an academic science department, is the norm. Each type of research setting comes with a set of implicit constraints on what is to be studied and who funds the work. The supporting institutions hold interests beyond the science itself that might bear on whether the results of the research become public, or whether they are applied, and in what way. Constraints on science can occur at high levels: as an item of national policy, such as a ban on use of federal funds for stem cell research outside a narrow confine, or a strong preference for funding select research areas in a federal agency, such as the National Science Foundation, or a private foundation, such as the Juvenile Diabetes Research Foundation.

In industrial laboratories, even quite basic research programs are expected to relate to the company's business interests. A few decades ago, General Electric's corporate research laboratory in Schenectady, New York;

IBM's T. J. Watson Research Center in Yorktown Heights, New York; Bell Laboratories in Murray Hill, New Jersey; or DuPont's Central Research Laboratory in Wilmington, Delaware, to name several outstanding examples, boasted basic research programs relatively uncoupled from short-term commercial application. In recent years, however, a more direct connection with the corporation's business interests has been expected in all corporate research environments.

In an academic department of a research university, constraints might come in the form of space limitations or lack of a necessary instrumental capability. A scientist may be hired into a department with the expectation that she will conduct her research in a given area, even though specific problem choices are hers alone. Sometimes a scientist would like to explore an area in which there is little interest at the time, so no funding is available. Work in areas of science that are out of favor could result in a lack of professional progress. In the social sciences there may be taboos against addressing some topics. These and various other kinds of pressures on faculty members and research administrators amount in the aggregate to inducements to focus on some areas of research and not others.

Autonomy as independence from external influences in choice of research problem and day-to-day research practice clearly applies to individuals. Similarly, powerful political forces can limit the autonomy of science as a social institution within a culture and thus limit its claims to both authority and moral authority. Writing in 1935, Robert Merton described the ways in which German science was forced to relinquish its traditional commitment to a scientific ethos, and summarized by writing: "The sentiments embodied in the ethos of science—characterized by such terms as intellectual honesty, integrity, organized skepticism, disinterestedness, impersonality—are outraged by the set of new sentiments which the State would impose in the sphere of scientific research" (1949, 596).

The case of German science was of course an extreme example of an imposition of state control. A similar quashing of scientific autonomy occurred in the Soviet Union during Stalin's reign and afterward, up to the advent of perestroika. But although German Nazism and Soviet Communism are particularly horrific examples, the autonomy of science as an institution is confined in every society by many kinds of political, social, and economic forces. And, inevitably, any constraints that subvert the claims of science to those of some other social sector diminish science's autonomy:

One sentiment which is assimilated by the scientist from the very outset of his training pertains to the purity of science. Science must not suffer itself to become the handmaiden of theology or economy or state. The function of this sentiment is likewise to preserve the autonomy of science. For if such extra-scientific criteria of the value of science as presumable consonance with religious doctrines or economic utility or political appropriateness are adopted, science becomes acceptable only in so far as it meets these criteria. In other words, as the pure science sentiment is eliminated, science becomes subject to the direct control of other institutional agencies and its place in society becomes increasingly uncertain. (Merton 1949, 597)

It is evident from this brief introduction that the autonomy of individual scientists and the science establishment as a whole, at all levels, are coupled to independence from at least some forms of outside influence. On the other hand, the influence of external forces is inevitable given that science is dependent on the recruitment of talented individuals into science careers, on financial and political support for research programs, and on a general societal acceptance of science as a beneficial agency in contemporary culture. The ways in which loss of autonomy may translate into loss of authority depend on many factors, which we will need to sort out as we proceed. Our study of the authority of science therefore involves the ways in which external forces act to constrain or alter the scope, direction, and character of scientific practice.

# TWO ○

The encounters with science that occur in the course of people's every-day lives shape their appraisals of science and how it fits into society. They also ultimately establish how much and what kinds of authority science can exercise. But science's place in contemporary society is also the product of historical development. To better understand the present relationships of science to society, it helps to know how things came to be the way they are. The entity that we think of as science does not have such a long history: perhaps four hundred years with respect to some of its aspects, much less with respect to others. It was formed from ideas about the natural world that are much older, however, dating back to the Ionian Greeks and the culture of Alexandria.

In this chapter we look at some of these developments with an eye toward what they reveal about authority. My aim here is not a detailed account of the historical development of Western science; I have neither the space nor the expertise for that. Rather, I am interested in the question of how science came to acquire authority, and in particular how that authority came at the expense of existing societal entities. As we move rapidly through history, it will be helpful to keep in mind a distinction between what is usually referred to as natural philosophy on the one hand, and science as an experimental and theoretical methodology on the other. The distinction is not always clear-cut; it engages both historical considerations as well as matters of practice. Epicurus, Aristotle, Ptolemy, Copernicus, Galileo, and Robert Boyle can all be counted as natural philosophers. The earlier philosophers in this group sought naturalistic explanations for the structure and form that they observed in the world, and for the changes they saw, but were not much given to experimental investigation. They were followed in time by those such as Galileo, who employed experimental studies to discover new phenomena and test hypotheses. Recourse to experimental studies was further developed in the course of the seventeenth century by Robert Boyle and others, who are associated with what is often referred to as the scientific revolution. There was a progression from musings and philosophizing toward a focus on experimental science grounded in planned, controlled experiments and data collection. Both the natural philosophers and the "scientists" who followed confronted issues of authority, though these were often of different kinds.[1]

## GREEK SCIENCE

Greek science may fairly be considered the fountainhead of modern science, at least as practiced in the Western world. It began in the fifth century B.C.E. with the ruminations of Ionian philosophers such as Thales, Anaximander, and others, who wondered about the ultimate building blocks of the world. The atomic theory, generally associated with Leucippus and Democritus, was an attempt to understand the visible world as built up from tiny, invisibly small units called atoms.

In the third century B.C.E., Epicurus developed the atomic theory further, as part of a larger philosophical outlook. Epicurean philosophy received its most elegant and extensive expression in about 50 B.C.E. in the poem *De rerum natura* by the Roman poet Lucretius. Apart from whatever merits the Epicurean view of nature might have as a physical model, his philosophy was a challenge to traditional theological views. He postulated a deterministic world of limitless boundaries, made up of atoms moving endlessly in a void. The Epicureans derided the idea that the gods determined the fates of humans. The Ionian philosophers began the profoundly important practice of asking questions about the natural world and attempting to answer them on the basis of observation and rational thought, rather than through recourse to dogma. Lucretius is quite explicit in contrasting Epicurean natural philosophy with religious beliefs:

1. The term "scientist" is relatively new; it was coined by William Whewell in about 1834, but was not widely used until some decades later.

Our starting point shall be this principle:
*Nothing at all is ever born from nothing*
*By the gods' will.* Ah, but men's minds are frightened
Because they see, on earth and in the heaven,
Many events whose causes are to them
Impossible to fix; so, they suppose,
The gods' will is the reason. As for us,
Once we have seen that *Nothing comes from nothing,*
We shall perceive with greater clarity
What we are looking for, whence each thing comes,
How things are caused, and no "gods' will" about it.
(Lucretius 1968, 24)

None of the great Athenian Greek philosophers accepted the Ionian atomic model of matter. Socrates, Plato's mentor, found it impossible to reconcile the Ionian mechanistic view of nature with his own preoccupations with the physical and spiritual issues of everyday human life. Plato followed Socrates in rejecting the Ionian approach. He thought in terms of mathematical models that could somehow reflect beauty and order. In *Timaeus*, Plato presents his scheme of the universe: Everything is to be understood in terms of the five possible regular solids. Four of the five regular solids represent the four elements of Empedocles: earth, water, air, and fire. The fifth solid, the dodecahedron (a solid having twelve plane faces), represented the universe as a whole.

Although Plato's scientific ideas were mostly nonstarters, his philosophy was highly influential for many centuries. The academy that he founded in Athens in 385 B.C.E. remained open until the beginning of the first century B.C.E. (Lindberg 1992, 73). The enduring influence of Plato's writings, particularly his accounts of Socrates' dialogues, lay in their appeal to a sense of place for personal and intuitive human experience in understanding the natural world. Aristotle, by contrast, went beyond Plato's range of metaphysics and mathematics to study every possible aspect of the natural world. Beginning in 335 B.C.E., Aristotle ran his own version of Plato's Academy, called the Lyceum. He wrote extensively on politics, metaphysics, ethics, logic, and science. Aristotle's inquiries into nature's workings were highly organized and conducted on a scale far exceeding anything that had gone before (Lindberg 1992, chap. 3). He compiled many volumes of observations, using an orderly procedure of defining the subject matter, commenting on what had been thought about it before, and then presenting his own arguments and solutions to the questions that had arisen. To judge from his writings, he was not a religious person, but he did allow that theology was study of the highest order. Nevertheless, his emphasis on investigating the physical world, and his naturalistic explanations, earned him the enmity of some, who brought charges of "impiety" against him. Whether as a result of conflicts over matters of religion or politics, Aristotle left Athens in 323 B.C.E. and retired to Chalcis, where he died within a year. Aristotle's experience with religious authority is an early example of what we see repeatedly in the history of Western science: In philosophical or scientific inquiry, changes in nature are attributed to natural causes, and natural explanations are sought, as opposed to a general reference to divine powers. Therefore the secular study of the natural world, even apart from whatever findings may eventuate, has the potential to create a conflict of authority with traditional religious beliefs and practices.

Aristotle's science ranged over many fields. He rejected the Epicurean idea that matter is formed from discrete units, each characteristic of its substance. He embraced Empedocles' model, in which the observable world is formed from four elements (earth, fire, air, and water). His interpretations of what he observed were based on the idea that there are four types of causation: formal, material, efficient, and final. Of these, only Aristotle's efficient cause relates to what we think of today as causation: some factor that drives a process or brings something into being. For example, sunlight causes trees to grow; it is an efficient cause for the growth of trees. His notion of final cause is based on the idea that processes in nature are driven to completion and regulated by movement toward a final end or state characteristic of each kind of substance. Thus, for example, when a stone falls from the hand to the ground, it does so because the final end of solid matter is to be at the center of the universe, which Aristotle believed was at or near the center of the earth. Gases (air), on the other hand, do not fall downward, because their final end is to be upward. Aristotle also believed that matter in the heavens obeyed different laws than did matter on the earth.

Aristotle excelled at close observation and classification. He amassed a remarkable storehouse of information on biological systems, both land and marine based. He classified animal species and arranged them into hierarchies in ways that are strikingly modern. He was quite interested in marine life; in studying dolphins he saw that they were mammalian, and classified them with land-based animals rather than fish. He also sensed that the animal kingdom could be classified in terms of a chain of progressive change, a precursor to a theory of evolution. His biological studies were his best work. He was not so successful in physics, where simply theorizing on the basis of everyday observation, without experimental testing, led him to many incorrect concepts. Ironically, his ideas in physics became very influential after his work assumed center stage in the medieval period.

## MEDIEVAL SCIENCE

Aristotle's writings narrowly escaped being lost to history altogether. Many of his manuscripts were found in a pit in Asia Minor around 80 B.C.E. by a Roman militia, brought to Rome, and recopied. They became very well recognized in the Arab world, and, through the writings of Arab-Islamic scholars such as Ibn-Rushd (Averroes, 1126–98), who lived in Spain, were widely known in the twelfth century. When the Arab provinces in Spain were regained by the Europeans, the Islamic storehouse of knowledge became available, and Aristotle's work was translated into Latin. His ways of approaching the study of nature, and the sheer breadth of his investigations, were a revelation to many twelfth- and thirteenth-century European scholars. Coincidentally, Averroes, who was the major vehicle for introducing Aristotelian ideas into Christian Europe, was himself persecuted by the conservative religious authorities of Islamic Spain. He held that one could reach the truth by two different pathways, philosophy or religion. As this amounted to a direct challenge to religious authority, he suffered imprisonment and exile.

The twelfth century ushered in a great revival of learning in Europe (Grant 2006). Greek philosophy had been rediscovered, and there was much interest in reconciling those writings with Christian doctrine. With respect to natural philosophy, Plato's writings, particularly his *Timaeus*, were the central texts. Increasingly, philosophers worked at developing cosmologies and physical theories based on natural causation. Scholars such as William of Conches argued that it was incumbent on the philosopher to look always for natural causes to explain what happens in the world, and have "recourse to God" only when natural explanations failed (Lindberg 1992, 200). But in a society dominated by the Catholic Church, this was dangerous ground. The natural philosopher could easily be seen as either denying the miraculous or contradicting scripture.

By the mid-thirteenth century, Aristotle's writings had gradually assumed prominence. The dominant philosophical movement in this time was scholasticism, a basic tenet of which held that faith and reason were conjoined—that there could be no inconsistency between them. Matters of faith were decided by the ecclesiastical authorities, who were assiduous in identifying and rooting out heresy. The church also oversaw the educational establishment in Europe. The growth of universities was an important development during the thirteenth century. They came about as schools associated with cathedrals expanded from just training clergy to attracting fee-paying students. To be attractive to students they needed to build up faculty from among scholarly clergy, some associated with orders such as the Dominicans or Benedictines. Certain legal immunities and privileges could be accorded to the scholars to induce them to remain at one place. While this was done locally at first, eventually the pope did bestow his benediction, as was done for Oxford in 1254, and Paris at about the same time.

Scholars at the universities were accorded a good deal of freedom, both in their studies and in what was taught. There was, however, one inescapable rule: what was said, preached, or written must be in accord with church doctrine, as defined by Rome. When questionable cases arose, a papal investigation was initiated, pretty much in the manner of a special prosecutor, to make a determination. It was in this environment that the writings of Aristotle aroused great interest when the Commentaries of Averroes on Aristotle, and translations of Aristotle's work itself, became available. Aristotle presented a challenge to the primacy of revealed truth by claiming that rational thought and observation form the path to truth, by asserting that the universe is eternal, and in his belief that the soul dies with the body. Because Aristotle's work was so widely studied, it was deemed important to reconcile it with church doctrine (Lindberg 1992, chap. 10). The first to offer a comprehensive interpretation of Aristotle's philosophy in terms acceptable to the church was Albert the Great (ca. 1200–1280). He is often regarded as the founder of Christian Aristotelianism. One of his students, Thomas Aquinas, a Dominican cleric at the University of Paris and the most important figure in scholastic philosophy, undertook in his *Summa contra Gentiles* to justify revealed religion. He went on in his masterwork, *Summa Theologiae*, to supply an Aristotelian substructure for Christian thought, arguing that faith and revelation were the bedrock of belief, and that all discoveries from reason could be interpreted in light of them.

The *Summa Theologiae* was completed in about 1273, just a year or so before Aquinas's death. Around the same time, Siger of Brabant, also teaching at Paris, followed Averroes's views on several matters that could be interpreted as having religious import (Wippel 1998). Notable among these was the idea that there is but one human intellect, existing apart from individual human beings. This would seem to deny the immortality of individual human souls. The flurry of new thought at Paris occasioned a papal investigation, led by bishop Stephen Tempier. Some 219 theses written by Paris students were condemned in 1277. The condemnations included not only works dealing with Aristotle, but also some of the writings of Aquinas himself, who was at that time already dead. Such condemnations, which could be appealed, doubtless had what today we refer to as a "chilling effect" on scholarship. Aquinas was canonized by Pope John XXII in 1323, however, which meant that his work was considered free of heresy. The scholastic accommodation of church doctrine with Aristotle's philosophy, as sanitized by Aquinas, was essentially complete.

What did all this have to do with the relationship of *science* to the church? Faith, after all, did not have much to say on the subject of physical science. It was rather on philosophical matters, such as the immortality of the soul, or whether the universe existed from eternity or was created, that conflict arose. When Copernicus published his great work in 1543, placing the sun rather than the earth at the center of the universe, it failed to occasion a substantial negative reaction from the Catholic Church. It was not until 1616 that *De revolutionibus* was placed on the Index of Forbidden Books, and only then pending correction of certain points. Protestants, however, who took a strictly literal line with respect to interpretation of biblical text, saw *De revolutionibus* as contradictory to the Bible. Of course the Roman Catholic Church also took the Bible as its doctrinal touchstone, but there was some precedent for flexible interpretation. For example, Saint Augustine, one of the hallowed early fathers of the church, had allowed that scripture could be interpreted figuratively with respect to the creation story (Augustine 1998, bk. 13). (In Augustine's time, however, the various documents that made up the scriptures had not yet been formed into the Bible as we know it today. Thus there was no single book to which a literal interpretation could be applied.)

A deeper and broader conflict was brewing between established religion and natural philosophy. Many factors, including the increasing availability of printed materials, were coming together to usher in a new age in which the divine would be increasingly separated from the exercise of authority (Rabb 2006). Copernicus's model carried with it the implication that the universe was much, much larger than had been supposed. Assuming that the earth revolved around the sun, in order to account for the astronomical observations of stars Copernicus had to place them at what seemed at the

time an unimaginably great distance, far beyond what Aristotle had estimated. The earth in its orbit around the sun became a mere pinpoint in the
vast heavens. Rather suddenly the universe came to be frighteningly large.
Could it be that God intended humanity to be the central feature of all this
immense creation? This idea was certainly not entertained in any explicit
way in the work by Copernicus, who was, after all, a canon of the Roman
Catholic Church, but it lay, as an ungerminated seed, in his work and that
of Johannes Kepler, who followed him.

## GALILEO

It was in the person of Galileo Galilei (1564–1642) that the growing
struggle for authority between established religious doctrine and natural
science came to the fore. Galileo occupies a unique position for many reasons, not least because he was the most important of the early experimen-
talists. Even in the sixteenth century, new observations of a new supernova
and of comets had begun to weaken Aristotle's authority on the nature
of the heavens. Galileo's observations and those of others on a supernova
that appeared in the sky in 1604 established that it was indeed in the stel-
lar realm, and not something closer, in the lunar heavens. Once properly
understood, these astronomical phenomena posed a contradiction for Aristotelian physics, according to which the heavens were immutable. Galileo's
successful 1604 observations of a supernova motivated him to improve
the telescope. In 1609 he began his famous observations of sunspots, the
moons of Jupiter, and the surface of Earth's moon. His speeches and writings on these subjects made him the most famous scientist in Italy.

Galileo's studies of mechanics, as well as his astronomical observations,
convinced him that Aristotle's physics and astronomy were mostly wrong.
He came to accept the Copernican heliocentric theory, which was in direct
opposition to the Ptolemaic model that Aristotle had promoted. His successes in promoting the heliocentric theory stimulated rebuttals among
the many Aristotelian natural philosophers. Aside from arguments that
the heliocentric model really failed to explain things better than the Ptolemaic model, Galileo's opponents argued that the Copernican model contradicted scripture and the church's interpretations of it. In fact, scriptural
support for a geocentric model is virtually nonexistent. While the Protestants had objected to Copernicus's theory on biblical grounds, it seems unlikely that anyone among the Catholic theologians had looked to scripture

for guidance on the matter until Galileo's case brought the issue to the fore. Galileo responded in his speeches and writings to the claims brought forth against the Copernican theory. In what proved to be a fateful move, he ventured into scriptural interpretation by attempting to show how scripture might be explained using Copernican theory.

In a letter addressed to the grand duchess Christina of Lorraine, but possibly intended for the cardinal Robert Bellarmine (Biagioli 2006), Galileo made a passionate plea for an accommodation of religious authority with that based on the use of natural reason. He begins with a lengthy defense of Copernican theory, and then proceeds to argue for a reasonably broad interpretation of the Bible on matters relating to the physical world:

> Hence I think that I may reasonably conclude that whenever the Bible has occasion to speak of any physical conclusion (especially those which are very abstruse and hard to understand), the rule has been observed of avoiding confusion in the minds of the common people which would render them contumacious toward the higher mysteries. Now the Bible, merely to condescend to popular capacity, has not hesitated to obscure some very important pronouncements . . .
>
> This being granted, I think that in discussions of physical problems we ought to begin not from the authority of scriptural passages but from sense-experiences and necessary demonstrations; for the holy Bible and the phenomena of nature proceed alike from the divine Word, the former as the dictate of the Holy Ghost and the latter as the observant executrix of God's commands. It is necessary for the Bible, in order to be accommodated to the understanding of every man, to speak many things which appear to differ from the absolute truth so far as the bare meaning of the words is concerned. But Nature, on the other hand, is inexorable and immutable; she never transgresses the laws imposed upon her, or cares a whit whether her abstruse reasons and methods of operation are understandable to men. For that reason it appears that nothing physical which sense-experience sets before our eyes, or which necessary demonstrations prove to us, ought to be called in question (much less condemned) upon the testimony of biblical passages which may have some different meaning beneath their words. (Galilei 1615)

This was bold writing indeed. Galileo suggested to the theologians that science might be in the best position to interpret certain aspects of creation, and that some passages of the Bible should be read in light of those interpretations. He was, in effect, proposing an authority based on natural reason and experiment. The letter was not, however, a manifesto of some sort, asserting the superiority of natural science over ecclesiastical authority. Rather, Galileo seemed bent on eradicating what he saw as false impediments to a mutual accommodation.

In any case, his effort failed. The widely circulated letter precipitated complaints against Galileo, charging that he had taken liberties in interpreting scripture. He was visited by the commissary general of the Holy Office of the Inquisition. When instructed to relinquish the opinion of Copernicus, Galileo acquiesced. He was a famous man, however, and a considerable stir had been created. In 1616, some seventy-three years after Copernicus's great work was published, the Congregation of the Index published the aforementioned proclamation, based on the judgment of an advisory panel of theologians. It judged the doctrine that the earth moves and the sun is motionless to be false and "altogether contrary to the holy scripture" (Allchin). In a letter from Cardinal Bellarmine, Galileo was enjoined to abandon completely the opinion that the sun stands still at the center of the universe and that the earth moves, and henceforth not to hold, teach, or defend such a position in any way whatever, either orally or in writing.

In 1623 Galileo published *The Assayer*, in which he further attacked Aristotle's authority and claimed that mathematical analysis is the way to understand nature. At this point he was in good standing with Pope Urban VIII. In 1632, however, he published *The Dialogues About Two Chief World Systems*, in which he presented claims in support of the Copernican theory and attempted to demonstrate that motion of the earth is possible. He had gone to great pains to obtain the necessary permissions from the relevant authorities, by putting the arguments for Copernican theory in the mouths of characters in a dialogue, and then in the end judging all of them deficient because they contradicted the teachings of the church. In employing such a tactic he misjudged his standing with the pope and other ecclesiastical authorities. What followed is an oft-told tale, at least in terms of its surface elements: Galileo's book was placed on the Index of Forbidden Books, and he was taken to Rome to stand trial. A special committee of cardinals determined that he had violated the injunction not to defend or teach the theory that the earth moves. He stood trial before the Inquisition, was found guilty of "suspected heresy," was forced to withdraw his support of Copernican theory, and, at the age of sixty-nine, was sentenced to home imprisonment.

The trial of Galileo is the most famous and controversial episode in the history of conflicts between science and religion. It is often pointed to as exemplary of the rigid, authoritarian stance of the religious establishment with respect to any challenges to its authority and hegemony. In this way of looking at things, Galileo was a martyr in the fight to liberate science from the smothering influence of religious dogmatism. On the other hand, extensive scholarly study of this episode, including all the social and political factors that obtained at the time, reveals a more complex and nuanced picture (Koestler 1959, pt. 5; Biagioli 1993, 2006; Margolis 1991; Shea and Artigas 2003; Finocchiaro 2005; McMullen 2005; Sobel 2000). This is not the place to review all of that; the decisive moment for our interests had come in 1616, when the theological consultants to the Congregation of the Index opined that the heliocentric theory contradicted Catholic faith. There followed a public edict that declared the heliocentric theory to be false. Whatever forces might have motivated the appointment of the advisory committee, its pronouncement and validation by the theologians stands as an unequivocal assertion of religious authority over certain beliefs and their expressions by individuals. In *The Dialogues* of 1632 Galileo argued for the primacy of the individual scientist's judgment over institutional authority. It would be a long time before that declaration transmuted into the authority of natural philosophy and science more generally—that is, before it became a matter of one form of *institutional* authority in opposition to another.

### THE BEGINNINGS OF MODERN SCIENCE

Although Galileo was forbidden to publish following his trial, he continued his scientific activities while under house imprisonment, and in the face of deteriorating health. In 1638, his new book, *Discourses and Mathematical Demonstrations Relating to Two New Sciences*, was smuggled out of Italy, printed, and widely distributed. The work summarized Galileo's experimental studies in mechanics and laid out the mathematical framework that he had developed to account for the observations. In important respects it was a departure from traditional ways of presenting ideas and conclusions about the natural world, a further step along the road to modern science.

René Descartes (1596–1650), roughly contemporary with Galileo, is famous for his relentless pursuit of absolute certainty, particularly as described in the *Meditations on First Philosophy*, published in 1641. His pursuit of philosophy was all-inclusive; it encompassed mathematics and the physical sciences, as well as ethics, anatomy, and what we would today term psychology. Building on the work of Galileo he essentially established the field of mathematical physics. He therefore represents in one person an important bridge between philosophy and science. Although Descartes was educated in a Jesuit college and considered himself a devout Catholic, he believed that it should be possible to separate reason and faith. This meant that he would deny many of the philosophical and even theological positions of the church while maintaining his Catholic faith. The church, on the other hand, did not necessarily approve of the philosopher, and Descartes felt the weight of ecclesiastical censure for much of his life. He moved to Holland in 1629 and lived in semi-seclusion there for twenty years, only occasionally returning to France.

By 1633 Descartes had ready a work on cosmology and physics, *Le Monde*, but he withdrew it from publication when he learned of Galileo's difficulties arising from rejection of the geocentric model of the universe. In 1637, however, he published three essays, entitled *Optics, Meteorology, and Geometry*, prefaced by an autobiographical introduction entitled *Discourse on the Method of Rightly Conducting One's Reason and Reaching the Truth in the Sciences*. This essay, which usually goes by the abbreviated name *Discourse on Method*, is Descartes' most notable. In it he writes of his 1633 treatise in light of Galileo's fate:

> Three years have now elapsed since I finished the treatise containing all these matters; and I was beginning to revise it, with the view to put it into the hands of a printer, when I learned that persons to whom I greatly defer, and whose authority over my actions is hardly less influential than is my own reason over my thoughts, had condemned a certain doctrine in physics, published a short time previously by another individual, to which I will not say that I adhered, but only that previously to their censure, I had observed in it nothing which I could imagine to be prejudicial either to religion or to the state, and nothing therefore that would have prevented me from giving expression to it in writing, if reason had persuaded me of its truth; and this led me to fear lest among my own doctrines likewise some one might be found in which I had departed from the truth, notwithstanding the great care I have always taken not to accord belief to new opinions of which I had not the most certain demonstrations,

and not to give expression to aught that might tend to the hurt of any one. (1637, 202)

Despite his attempts to pacify religious authority, Descartes did not escape censure, albeit posthumously. His work was officially placed on the Index of Forbidden Books in 1663 by the church in Rome, thirteen years after his death.

The Galileo affair clearly put on hold Descartes' plans for disseminating his work; other natural philosophers doubtless felt similar constraints. Descartes and Galileo represented the vanguard of a new way of doing science; while it drew on the resources of medieval science, it departed from the past not only in methodology, but also in its firm rejection of Aristotelian metaphysics (Lindberg 1992, chap. 14). As science began to take shape as an entity in its own right, the stage was set for increasingly sharp conflicts of authority with the church. At this time France was becoming a focal point for philosophers and theologians interested in natural philosophy. Pierre Gassendi, a French cleric and theologian, published in 1642 an exposition of Galileo's mechanics. Gassendi had earlier become thoroughly dissatisfied with Aristotelianism. He undertook a long-term project of reinterpreting Epicureanism in terms that would be acceptable to orthodox Christianity. His massive work, *Syntagma philosophicum*, covers logic, physics, ethics, and political philosophy. The empirical approach to gaining knowledge that he advocated was influential with later natural philosophers such as Robert Boyle and John Locke.

Marin Mersenne was also a French cleric, a member of the religious order of the Minims, whose members devoted themselves to prayer and study. He lived in Paris for nearly his entire life. Although he was an accomplished mathematician and did important work with prime numbers, he is remembered also as a facilitator of interactions between natural philosophers. It was largely through Mersenne that Galileo's lectures on mechanics became known throughout Europe. He organized meetings of the leading philosophers of the time, and was a prodigious correspondent. He assisted in publication of Descartes' *Discourse on Method* in 1637 and defended both Descartes and Galileo against the criticisms of theologians. He also collected the criticisms of several philosophers of Descartes' *Meditations on First Philosophy*; these and Descartes' rebuttals were published in 1644 along with the second edition of the work. He organized conferences that brought together such figures as Gassendi and Blaise Pascal, and which included at times the English philosopher Thomas Hobbes. Both Gassendi and Mersenne, as members of religious orders, placed orthodox Christian beliefs ahead of all others. Their concern was always to accommodate the study of the natural world to those beliefs. In his exposition of Epicureanism, Gassendi made sure that what he kept of the Greek philosophy was consistent with church doctrine. Mersenne maintained an active correspondence with certain Protestant philosophers, in which he strove to overcome doctrinal differences.

Much of Descartes' writing, as well as that of Gassendi and Mersenne, can be seen as an attempt to rebut skepticism, an ancient Greek philosophical outlook that enjoyed a revival beginning in the mid-sixteenth century. These skeptics held that no proposition could be held as absolutely certain, that there was always some room for doubt. As a result, they felt it best to avoid subscribing to any dogmatic belief; instead, one should simply go along with prevailing norms. Theodore Rabb (2006) contends that the rise of skepticism, expressed in the writings of people such as Michel de Montaigne, can be seen as a reaction to the uncertainties created by the speculations of natural philosophers. The gradual replacement of the thoughts of antiquity by new visions of the world seemed to leave confidence and truth beyond reach. Seen from a religious perspective, skepticism had the potential for creating doubt about revealed religious truth. To counter universal claims of skepticism, Mersenne had recourse to the mathematical sciences, conventionally treated as an exemplar of certain knowledge. By applying mathematical methods to the modeling of observations, the truth of ordinary sense data could be established. Gassendi responded to the challenge of skepticism in a different way. He rejected Aristotle's ideal of demonstrative certain knowledge as an epistemological criterion, advocating instead that we can obtain *probable* knowledge of things based on the evidences of our senses. He argued, in effect, that it is not necessary to obtain certain knowledge to make progress in understanding nature. Gassendi was among the first empiricists among scientific philosophers.

The English philosopher Thomas Hobbes was strongly influenced by the French school of philosophy. Like Mersenne and Gassendi, he defended natural philosophy against skepticism. Hobbes left England in 1642 for political reasons and spent many years on the Continent. He attached himself to the Parisian network centered on Mersenne, and became well-known and respected in that circle. His writings reflected a strongly materialist orientation, as well as a strong antipathy toward religious authority and the scholastic philosophy that in his mind had accepted and defended that authority. He argued that human judgment, influenced as it is by artful suasion, self-interest, pains, and pleasures, is an unreliable source of wisdom. In his 1651 masterwork, *Leviathan*, he speaks of science as "the knowledge of consequences" (43), and as the only means for understanding the present and forecasting the future. Hobbes's métier was politics. His abiding concerns had mostly to do with how to create a just and stable society. He saw science as a means toward that end. Science directed toward the natural world had the capacity to improve human life through a better understanding and control of nature. Analogously, scientific principles should be sought for governing human affairs. Although Hobbes did not contribute much of direct value to natural philosophy, he did ascribe to science an authority to lay superior claims to knowledge and understanding.

In challenging radical skepticism, these philosophers built one of the foundations for authority of the science that was to come: the idea that it *is* possible to form and entertain meaningful propositions about the physical world, and to achieve a high level of confidence in conclusions that we may draw from its study. In addition, by testing the boundaries of what was acceptable to ecclesiastical authority, whether Roman Catholic or Protestant, Galileo and those who followed him created a space in which natural philosophy could exert an intellectual authority that it had previously lacked.

### EXPERIMENTAL SCIENCE

The English philosopher and statesman Francis Bacon (1561–1626) led the charge in English culture against Aristotelian logical analysis in favor of inductive empiricism in studies of nature. He contended that nature could be understood only by systematic experimentation. He favored an inductive approach in which rules or laws were, in essence, a summary statement of accumulated observations. Bacon was derisive of the system of teaching in the colleges, and called for the establishment of new social structures that would more effectively advance learning (Bacon 1605). He was outspoken in his defense of natural philosophy with respect to religious authority:

> Neither is it to be forgotten that in every age Natural Philosophy has had a troublesome adversary and hard to deal with; namely, superstition and the blind and immoderate zeal of religion. . . . But if the matter be truly considered, natural philosophy is after the word of

God at once the surest medicine against superstition, and she is rightly given to religion as her most faithful handmaid, since the one displays the will of God, the other his power. . . . Meanwhile it is not surprising if the growth of Natural Philosophy is checked, when religion, the thing which has the most power over men's minds, has by the simpleness and incautious zeal of certain persons been drawn to take part against her. (Bacon 1620, 89)

He also urged a new level of receptivity toward new knowledge, a recognition that the ancients could not have been aware of much that was now knowable: "By the distant voyages and travels which have become frequent in our times, many things have been laid open and discovered which may let in new light upon philosophy. And surely it would be disgraceful if, while the regions of the material globe . . . have in our time been laid widely open and revealed, the intellectual globe should remain shut up with the narrow limits of old discoveries" (1620, 82).

William Harvey (1578–1657), an English physician who received his training at the famous medical school in Padua, Italy, was well acquainted with Bacon, and was one of the first to successfully apply Baconian ideas of natural philosophy. Upon returning to England after his medical training, he married the daughter of Queen Elizabeth's physician. He was well connected and therefore had some leisure time to devote to scientific studies. He developed an interest in the circulation of the blood, a subject on which little advance had been made since the time of Galen in the second century C.E. Harvey concluded on the basis of extensive experimental work that the heart pumped blood into the arteries, and that blood circulated via return of the blood through the veins. His model clashed with the accepted idea, based largely on Galen, that the venous and arterial blood systems were separate, and that there was no general circulation. He began to lecture on his theory in the College of Physicians in 1616, but his work, *Exercitatio Anatomica de Motu Cordis et Sanguinis in Animalibus*, commonly referred to as simply *De Moto Cordis*, was published in Germany only in 1628. Harvey had this to say of his research, which is deemed today to be one of the great scientific studies of the seventeenth century: "[It] is of so novel and unheard-of character, that I not only fear injury to myself from the envy of a few, but I tremble lest I have mankind at large for my enemies, so much does wont and custom, that has become another nature, and doctrine once sown and that has struck deep root and rested from antiquity, influence all men" (Harvey 1628).

Harvey fully recognized the power and authority of the royal house, and his book begins with a letter to "The Most Illustrious and Indomitable Prince, Charles, King of Great Britain, France, and Ireland, Defender of the Faith," which says, in part: "Accept therefore, with your wonted clemency, I most humbly beseech you, illustrious Prince, this, my new Treatise on the Heart; you, who are yourself the new light of this age, and indeed its very heart; a Prince abounding in virtue and in grace, and to whom we gladly refer all the blessings which England enjoys, all the pleasure we have in our lives." The cool reception given to Harvey's discoveries illustrates well the new science's difficulty in overcoming entrenched beliefs. He correctly anticipated that his work would be attacked, although it was eventually accepted during his lifetime.

Bacon also exerted a powerful influence on Robert Boyle (1627–1691), unquestionably one of the leading intellectual figures of the seventeenth century. Boyle's importance for us is that he was perhaps the most influential practitioner of the newly developing experimental approach to philosophy. Boyle's father, Richard Boyle, the first Earl of Cork, had come from non-gentlemanly stock, amassed a fortune through a series of unscrupulous moves, and attained a peerage through his money. Despite his father's various legal difficulties, Robert as a young man was the inheritor of a substantial and quite secure fortune. His father had been assiduous in assuring Robert's education, through school at Eton and particularly through the services of a private tutor.

Robert Boyle had no interest in acquiring a peerage. He was thoughtful, introspective, and deeply religious. In about 1656 he took up residence at Oxford. By this time he had already developed an interest in natural philosophy, and had joined a circle of gentlemen interested in promoting what they referred to as "physico-mathematical experimental learning." In 1660 this group, numbering a dozen, officially founded a college, and began weekly meetings to witness experiments and discuss scientific topics. It is reasonable to suppose that Francis Bacon's writings had encouraged them toward this move. The founding group decided to invite an additional forty members, thirty-five of whom accepted the invitation. Of this number, nineteen were considered men of science, while the other sixteen included statesmen, soldiers, antiquarians, administrators, and one or two literary men. The venture received the approval of the king, Charles II of England, and within a few years had the title "The Royal Society of London for Improving Natural Knowledge."

Steven Shapin (1994) has emphasized the significance of social standing among those who in mid-seventeenth-century England both carried out or supervised the conduct of experimental science and controlled the manner in which it became public knowledge. All members of the Royal Society were "gentlemen," and many were well-off financially. Among his compatriots Boyle was the most adept and ardent about experimental work; he quickly became a paragon of the experimental philosopher. He published several volumes, including the enormously influential *New Experiments Physico-Mechanical, Touching the Spring of Air and its Effects* in 1660, and *The Skeptical Chymist* in 1661. The development of English experimental science proceeded largely without conflicts with religious authority. In company with most of the men of his circle, Robert Boyle was conventionally religious. For example, he embraced the account of creation in the first two chapters of Genesis as literally true, and accepted the date of 4004 B.C.E. for creation propounded by James Ussher, archbishop of Armagh. He wrote extensively about how God must have gone about creating the world in just six day's time, whether natural laws could exist, how humans compare with the angels, and many similar topics. The potential for conflict with religious authority had, in his time and place, retreated into the background. There were, however, other ways in which the newly emerging entity "science" evoked issues of authority.

Shapin's account of Boyle and his circle of natural philosophers has several points of interest. First, all the members of the circle were "gentlemen," a term that denoted both wealth and "breeding"; this meant that they were presumed to have leisure to pursue intellectual interests, without having to concern themselves about whether there was financial gain to be had from whatever might be learned. (In fact, some members of the Royal Society were not wealthy, and they struggled to maintain their gentlemanly status. Stephen Inwood's 2002 biography of Robert Hooke describes the social relationships within the Royal Society, and Hooke's important roles in particular, in a somewhat different light from Shapin.) Second, English gentlemen were assumed among themselves to be entirely truthful, above any temptations to mendacity. Third, they wished to be regarded as disinterested scholars, pursuing their studies without any thoughts of self-promotion. Shapin notes that Boyle's texts were advertised as printed forms of letters written to particular individuals. This was a ploy on Boyle's part; he did not want to appear to be seeking notoriety or in any other way be viewed as self-aggrandizing. Science was not a workaday profession, but rather an elective activity of men who had the means to choose how they spent their time. These assumptions about the scientists of the Royal Society as individuals, and about their activities and claims, are the origins of a tradition of scientific authority vested in the qualities of those who publicly describe their scientific work in both demonstrations and writing. As the social standings of those practicing science changed, the idea of a scientific authority deriving from the personal characteristics of the scientist became less central, though the personal charisma of individual scientists of great eminence continued to be important.

The English natural philosophers, particularly Boyle, followed the Baconian program in designing equipment and experimental procedures to generate new data. For example, Boyle and Robert Hooke carried out extensive studies of the properties of gases, employing an air pump invented by Hooke. Among their findings, they established that a gas confined in a closed tube could be compressed, and that the volume of the gas related to the pressure acting on it. Their results showed that the volume of the gas decreased in proportion to the applied pressure, over about a fourfold range of pressure change. Boyle was reluctant, however, to offer models for such phenomena. Though pressed to offer an explanation for the results, Boyle declined to do so, referring only to the "spring of the air." His strongly empirical approach to experimental science, coupled with a reluctance to speculate on the origins of the effects he observed, ensured that his pious religiosity would not come into conflict with his scientific work. More important, in terms of the public face of science, it further shifted the emphasis away from the speculative characteristic of Greek science toward active investigation of the world's properties.

## THE EXPANDING SOCIAL STRUCTURE OF SCIENCE

By the mid-seventeenth century the world of science was growing. The membership of the Royal Society increased as new members were elected, including those from the Continent. As it expanded, particularly with the inclusion of foreign members, the criteria for membership were weighed less toward gentlemanly credentials and more toward scientific accomplishment and reputation for rectitude. Organizations similar to the Royal Society were established in other European countries. Most notably, the Royal Academy of Sciences in Paris was founded in 1666. Communications of individual natural philosophers with one another and with the societies increased. Publication of scientific results became more regularized. In 1665 the first issue of the *Philosophical Transactions* was published, and Henry Oldenburg, the Royal Society's secretary, served as editor. In time it became the *Philosophical Transactions of the Royal Society*, which is still published to this day.

As the number of people actively pursuing scientific interests increased, and as channels of communication became more extensively deployed, science increasingly took on characteristics common to all social organizations. But aspects peculiar to science did begin to surface, among them the emergence of conflicts among scientists who appeared with differing reports on what was presumably the same set of phenomena. For example, Shapin gives an account of a dispute between two astronomers, Adrien Auzot, a well-respected natural philosopher of Paris, and Johannes Hevelius, a wealthy brewer and senator of Danzig as well as an eminent astronomer (1994, 267–91). Hevelius was made a fellow of the Royal Society in 1664, and Auzot in 1666. Their dispute had to do with the proper placement of the so-called comet of 1664 on the night of February 8, 1665. After much dithering, a panel of distinguished philosophers and astronomers, including Boyle, Hooke, and Christopher Wren, came down on the side of Auzot, although it did provide Hevelius an opportunity to acknowledge gracefully that he had made an error. In 1672, when Isaac Newton, newly elected fellow of the Royal Society, published his first scientific paper on light and color in the *Philosophical Transactions of the Royal Society*, Robert Hooke and Christian Huygens (a Dutch physicist and astronomer) challenged Newton's attempt to prove by experimental means that light consisted of corpuscular motion rather than waves. Newton reacted with a rage that could fairly be called irrational, and the incident triggered a long-standing animosity between him and Hooke.

These early examples illustrate a pervasive characteristic of scientific research: a passionate regard for being right, and—just as important—for being thought right. Whether the matter concerns observations, a contested interpretation of data, or a theoretical construct formed to account for observations, scientists can be very disputatious. They are also driven to be recognized as original—to be credited as the source of novel insights into nature—whether in the realm of experiment or theory. Robert Hooke was notoriously polemical and became embroiled in many arguments over priority of discovery (Inwood 2002). Even Robert Boyle frequently complained, especially in his later years, that his experimental work was being expropriated by others and presented as their own. In the course of his research life he became continually more careful about how his writings were produced and circulated. This apparently was not in the interests of financial gain, as he had no need for additional income; rather, as modest and retiring as he was, he nevertheless valued his reputation as the premier experimental scientist of his time. Newton furnishes yet another example. In the latter part of his life he was arguably the most famous scientist alive. He was elected president of the Royal Society in 1703 and reelected every year until his death in 1727. In 1705 he was knighted by Queen Anne, the first scientist to be so honored for his work. Despite all this success, the last years of his life were tormented by his contention with Gottfried Leibniz over who should receive credit for invention of the calculus.

Much more could be said about social considerations largely internal to science itself. They raise issues of authority, such as who has the authority to resolve disputes among scientists, or to adjudicate competing claims for priority. They also prompt the question of what factors determine the authority of a given scientist among his or her peers. For example, after Newton the wave theory of light did not return to favor until the nineteenth century. Newton's scientific authority was so great that many proposals conflicting with his interpretations were dismissed out of hand. The processes by which authority is established and maintained are a key element in science's internal workings; they are also important for establishing science's authority in society at large.

We have identified four major aspects of authority that surfaced in the history of early science. First, in their emphasis on rational analysis of observations of nature, the Greek natural philosophers created a new mode of thought that challenged traditional, myth-oriented accounts. Their successes in systematizing and explaining many aspects of the natural world imbued their claims with authority. Further, they argued that it was possible to arrive at irrefutable truths through logical analysis, thus establishing a new basis for authority.

Second, early modern science had its beginnings in a culture based on religious authority. Beginning in the late Middle Ages, new models of the natural world, occasioned by new experimental findings or by new ways of viewing data already on hand, created challenges to the authority of a Christianized version of Aristotle's natural philosophy. Ecclesiastical authority, both Roman Catholic and Protestant, was marshaled at times to quench ideas and movements deemed in conflict with the church's interpretations of scriptural writings. Even though most of the early practitioners of science took it for granted that conclusions drawn from natural philosophy were subordinate to their religious beliefs, scientific authority gained ground. It gradually created facts about the world and its workings that simply would not fit within the boundaries of traditional beliefs. Ecclesiastical doctrine, and even scriptural interpretation, underwent change in response to the implications of new scientific findings.

Third, in response to radical skepticism, a challenge to authority of any kind, the natural philosophers argued that the study of nature could lead to empirical knowledge known with sufficiently high probability that it could count as useful knowledge. In doing so they laid the groundwork for scientific authority as we know it today. Lastly, as natural philosophy in its traditional forms metamorphosed into "science" through recruitment of more practitioners with more varied social and technical backgrounds, science became a social entity with its own set of norms of personal conduct, standards of investigation and data gathering, means of communication, and ways of presenting itself to the larger society. The canons of disinterestedness, truthfulness, openness, objectivity, and absence of self-aggrandizement subscribed to by those that formed the first scientific societies became articles of faith among subsequent practitioners of science. To the extent that they were acceded to, both within and outside the science community, they sustained claims of authority for science against other outlooks, such as religion, law, and social theory.

The Enlightenment scientists made very important discoveries that set the stage for much further advancement in understanding nature, but they were not much concerned with applying their ideas in practical ways. For example, William Harvey's discovery of the circulation of the blood did not result in materially changed medical practices for many decades. To be sure, there were also exceptions. For example, Robert Hooke was a great inventor (Inwood 2002). He invented a better microscope than had previously been available and used it to create his 1665 masterpiece, *Micrographia, or Some Physiological Descriptions of Minute Bodies Made by Magnifying Glasses.* In that volume he reported observations that vastly extended what was known about life-forms. In addition, he invented the anchor escapement and the balance spring, essential elements of greatly improved clocks; meteorological instruments such as the anemometer and hygrometer; and many tools for measurement. But Hooke, one of the great scientists of the seventeenth century, was exceptional; applications of their scientific understandings were not uppermost in the minds of most early modern scientists. That situation changed over time. By the eighteenth century, many scientists were involved in the study of practical problems with an eye toward their solutions. For example, Antoine Lavoisier, a member of the French Academy who lived in Paris, wrote papers on lighting and potable water for the city and assisted with other public works projects.

I end this chapter by fast-forwarding to briefly review the career of Louis Pasteur. This famous figure furnishes a prime example of how charismatic authority could be created and exercised in nineteenth-century science. Pasteur's career highlights the application of scientific findings to practical problems. Pasteur undertook the challenge of addressing societal issues, and the results of his endeavors had important consequences for his and science's authority.

### PASTEUR: SCIENCE IN THE WORLD

Louis Pasteur was born in 1822, in Dole, France, the son of a poorly educated tanner. As he grew up in the nearby town of Arbois, he showed sufficient precociousness that the headmaster of the college at Arbois encouraged him to try for the most prestigious French university, the École normale supérieure in Paris. He was admitted and in due course obtained his doctorate in chemistry in the laboratory of Antoine Ballard. In pursuing his doctoral research he made his famous observation that crystals of racemic tartaric acid observed under the microscope existed in two asymmetric forms. By manually separating these two forms he showed that one of the two forms was identical to a form of tartaric acid commonly found in the sediments of fermenting wines. The other form, he postulated, was the same compound but in a different three-dimensional form.

Some years later Pasteur served a professor of chemistry at the Faculty of Sciences in Lille, France, a largely industrial town with a number of distilleries and factories. Pasteur took an active interest in the problems encountered in the factories, and was quick to respond to the entreaties of one M. Bigo, who was having trouble manufacturing alcohol by fermentation of beetroot. The fermentations sometimes produced no alcohol, but instead formed undesirable products such as lactic acid. The conventional wisdom at the time was that fermentation was a straightforward chemical process in which sugars were converted to alcohol, carbon dioxide, and water. The fact that yeast cells were found in the fermenting vats was deemed incidental to the fermentation process. Pasteur studied the problem at length and established that yeast was in fact essential to the formation of alcohol, and that the process went awry when contaminating microorganisms turned the fermentations sour. In the course of a research program extending over several years, Pasteur identified and characterized specific microorganisms responsible for the desired and undesired fermentations in productions of beer, wine, and vinegar. He also showed that heating the beer, wine, milk, or other microbe-containing solution to a moderately high temperature for a few minutes could quench the undesirable fermentations. He thus originated pasteurization, a process that has been of incalculable value to human welfare.

During this time, "spontaneous generation" was still a subject of debate in scientific circles. Spontaneous generation is the theory that nonliving substances can give rise to living organisms in some direct way. For example, it was thought that meat, left standing in the air, could give rise to flies. This particular instance of folk theory was put to rest in 1668 by Francesco Redi, an Italian physician who showed that whereas meat exposed to air in an open jar gave rise to flies and maggots, meat placed in similar conditions but in a tightly closed jar did not. Redi concluded that the flies entering the open jar had deposited eggs on the exposed meat, and that maggots that appeared had hatched from the eggs rather than just appearing spontaneously as living organisms.

Although Redi's experiment ruled out the idea that larger entities such as flies could arise through spontaneous generation, it was still widely held that the very small entities revealed under the microscope had come into existence spontaneously from the substances in which they were found. The debate was very much at the forefront when the Paris Academy of Sciences offered in 1862 a prize to be awarded for any experiments that would definitively resolve the conflict. This was just Pasteur's sort of challenge. In 1864 he received the Academy's prize by putting forward experiments that definitively ruled out spontaneous generation. He boiled several samples of soup, but each was then treated in a different way. He allowed all the samples to be exposed to air, but in one case the opening to the container was blocked with a cotton plug, which Pasteur had shown did not admit the passage of bacteria. Another flask had a specially designed swan-neck (or S-shaped) opening, which Pasteur believed would not admit the passage of bacteria because they would settle on the walls of the initial portion of the neck. As he predicted, soup in the flasks that had the cotton plug or swan-neck opening did not spoil, even though they were open to the air, whereas the soup in a straight-necked flask, without a cotton plug, did spoil. What is more, when the cotton plug was removed, the soup in that container spoiled, as did the soup in the swan-neck flask when the neck was cut off to eliminate the S-shaped bend. All the soups had been boiled to the same extent. Pasteur's conclusion was presented quite dramatically: "I have removed life, for life is the germ and the germ is Life. Never will the doctrine of spontaneous generation recover from the fatal blow that this simple experiment delivers to it. No, there is now no circumstance known in which it can be affirmed that microscopic beings came into the world without germs, without parents similar to themselves" (Geison 1995, 120).

The debate over spontaneous generation was largely fought between Pasteur and Felix Pouchet, a French naturalist. It was a high-profile contest with religious and political overtones. Gerald Geison (1995, Chap. 5) has described in detail how Pasteur managed the public face of the debate to his advantage. For example, his lecture of April 7, 1864, dealing with spontaneous generation and its religio-philosophical implications, was attended by the glitterati of Parisian society, and Pasteur gave them a brilliantly staged performance. A strong focus on his public image became a hallmark of Pasteur's scientific career (although that image sometimes masked scientific work that fell short of his claims for it).

Pasteur built his career by deliberately directing his interests toward solving problems of importance in both industry and health. He had a great eye for the simple, strongly inferential experiment, one that excludes one or more alternative hypotheses. Because he constructively accumulated understanding of the underlying mechanisms operating in the systems he studied, each of his endeavors informed those that followed. For example, he was asked by the Department of Agriculture to head a commission to see what could be learned about a new disease devastating the French silk industry. He identified two distinct microorganisms as responsible for incapacitating the silkworms, and showed the silkworm farmers how to select worms that were free of the diseases. These procedures of selection were sufficient to restore the silk industry in Europe. For Pasteur the study brought to light new insights into how the diseases spread, and the conditions that made for resistance to infection. This acquired knowledge formed the basis of his later and more celebrated work in controlling disease.

Pasteur's work on contamination of wine and beer by airborne yeasts made it evident that bacteria are ubiquitous in our surroundings. This understanding stimulated the physician Joseph Lister in Glasgow to apply carbolic acid to open wounds in order to kill microorganisms at the wound site. He also systematically sterilized instruments, bandages, and the working surfaces in his operating quarters, thereby greatly reducing infections. But the medical world was slow to change its way of doing things, or to recognize that many important diseases, such as cholera, diphtheria, scarlet fever, and childbirth fever, were caused by microorganisms. Pasteur, whose reputation had become great, urged physicians to take precautions. In a speech before the Academy of Medicine he urged adoption of the new procedures. He recommended that surgeons use perfectly clean instruments, scrub their hands with the greatest care, and use only bandages and other materials that had previously been exposed to high temperatures (Debré 1994, chap. 10). By instructing the members of the august Academy as to what they *should* do in the course of surgical work, Pasteur was exercising *moral* authority. Despite his exalted reputation as a scientist, however, his moral authority with the medical establishment was quite limited. The members of the distinguished Academy of Medicine, a generally self-satisfied lot, did not graciously accept advice from someone who, though a newly elected member, was outside their circle.

Pasteur was working on containment of chicken cholera when he and his coworkers first identified the cholera bacillus, and then learned to grow it in pure culture. When injected with the cultures, chickens invariably died within two days. Pasteur and his coworkers then learned how to attenuate the virulence of the cultures by a combination of exposure to air and warmer temperatures. By manufacturing attenuated cultures of chicken cholera in a controlled manner, Pasteur routinely prevented chicken cholera in vaccinated chickens. Using similar methods, he then produced a vaccine against anthrax. There was a fierce competition among French scientists not only to develop a reliable procedure for preparing an effective vaccine, but also to be seen as having definitively solved the problem. Pasteur was not above using a variety of resources at his command to denigrate in both scientific and public sectors the contributions of his competitors (Geison 1995, chap. 6). His behavior in this instance is but one more example of the competitive drive for priority that permeates scientific research (Bucchi 1998).

To add to Pasteur's concerns over the difficulties in preparing a reliably effective vaccine, Hippolyte Rossignol, a veterinary surgeon who did not accept Pasteur's germ theory, openly questioned his authority. Rossignol challenged Pasteur to carry out an independent public trial: twenty-five sheep were to be vaccinated and then exposed to anthrax, while another twenty-five were to be given just the exposure to anthrax. The test was to take place at a farm near the commune of Pouilly-le-Fort, south of Paris. Although his trials with the anthrax vaccine were by no means complete, Pasteur accepted the challenge. The publicity attending the event was intense. The London *Times* and the French papers covered the event with daily bulletins. Happily for Pasteur, the trial was a complete success. Within a few days of inoculation all twenty-five of the control sheep were either dead or very ill, whereas the twenty-five inoculated sheep were alive and healthy. The episode is more nuanced than I have indicated, but these broad features serve for our purposes. It was a great public triumph for Pasteur, captured in this somewhat ironic passage from Gerald Geison's volume: "In no small part, Pasteur succeeded because of his flair for the dramatic gesture and his talent for self-promotion" (1995, 268).

According to some historians, it was the fame of the anthrax vaccine experiment that engendered the public's faith in the science of microbes. This one dramatic demonstration persuaded many nonscientists to accept the authority of science as to the existence of normally invisible entities. Within a decade, a total of 3.5 million sheep and half a million cattle had been vaccinated against anthrax, and the total mortality rate was less than 1 percent, all of which meant an enormous savings to the French economy. Pasteur went on to identify and isolate the microbes responsible for many other animal and human diseases, including pneumonia and childbirth fever. But there was yet to come one further scientific accomplishment that would vault him to still-greater notoriety, making him a national hero.

Rabies, a viral disease contracted by humans upon being bitten by a rabid animal, was one of the most dreaded afflictions in the public mind. Though not all that commonly contracted, it had an awful reputation because it was thought to be nearly always fatal and to cause the victim great agony. The key to a vaccine was to devise an agent that could confer protection before the rabies virus traveled from the site of the bite (from a rabid animal) to the spinal cord, and then to the brain. Pasteur and his colleague Joseph Roux were aware that success in finding a vaccine for rabies would be seen as a great triumph; Edward Jenner's vaccine for smallpox was the only one known for a human disease. Pasteur and Roux were able to develop a vaccination regimen for dogs that consisted of daily injections, over about twelve days, of increasingly potent material formed from the spinal cord of rabid rabbits. They reported in 1885 that they had made dogs refractory to rabies through such a vaccination procedure, but there was fear that the procedure might not work with a human subject. In the following year, while trials with dogs were still in progress, nine-year-old

Joseph Meister and his mother appeared at Pasteur's door. The child had been bitten repeatedly by a rabid dog, and it appeared virtually certain that he would contract rabies. After consultation with medical authorities, and despite misgivings about using the vaccine on a human, Pasteur consented to treat the youth. The boy made an excellent recovery, and remained in good health for the remainder of his life.

It was not long before a second victim begged for vaccination and was successfully treated. Word spread everywhere, and victims of rabid animal bites poured into his laboratory. The attendant publicity was immense; Pasteur became a heroic legend in his lifetime. The Pasteur Institute, funded by governmental and private money, was built in Paris to deal with rabies and other diseases associated with Pasteur's work. He became the founding director of the Institute in 1888, although his health was failing. He had suffered a stroke in 1868, which left him with impaired speech and gait, and which made it difficult for him to directly take part in the day-to-day operations of his laboratory. Indeed, these physical impairments worked to create a more sympathetic light over an intensely combative and ambitious man (Geison 1995, 270). He suffered additional strokes in his last years and died in 1895. After a funeral attended by thousands of people, his remains were interred for a time in the Cathedral of Notre Dame before being transferred to a permanent crypt at the Pasteur Institute.

Pasteur's story is that of an individual scientist who, through his scientific research, made wide-ranging, important contributions to society by improving industrial processes, finding solutions to problems in agriculture and animal husbandry, and developing protective measures against dreaded diseases. On many occasions during his career, his authority as a scientist was called into question, such as when he was accused of failing to give a full and open account of his experimental methods. These were serious charges, because in science authority is created by producing experimental evidence in support of scientific claims. These disputations might have remained an issue internal to the scientific community but for the economic, social, or medical implications of Pasteur's work. He was skillful in playing these aspects of his work for all that they were worth, and in the process drew the interest of a large public outside science. The showdown with Rossignol in connection with the anthrax vaccine is a prime example. It took on the character of a major public event, in large measure because Rossignol himself wanted to make it so, confident that he would expose Pasteur as a fraud. By the time that Pasteur developed his rabies vaccine, however, his reputation generated all the public notice he could have wanted. Success with the rabies vaccine ensured his entry into the select circle of France's heroes.

It is clear that Pasteur had charismatic authority in society at large. This was grounded not only in what he could convincingly claim to know, but also in how that knowledge was applied to problems of great economic and medical importance. The fact that his knowledge claims could be validated through their successful application to important social problems imbued them with great authoritativeness with both scientists and the lay public. Pasteur also had great charismatic authority because he actively strove to be seen as a disinterested scientist searching for truth without expectation of personal fame or remuneration. Although that characterization of him has been questioned, what matters for evaluating Pasteur's authority is how he was perceived by society at large. In the eyes of nearly everyone, he was a heroic figure.

Finally, one might question what uses Pasteur made of his charismatic authority to exercise *moral* authority. I gave an example above, in which he urged surgeons to ensure more aseptic conditions in their medical practices. Of course, in advising silkworm farmers on how to minimize damage from silkworm diseases, or sheep farmers on ways to minimize the spread of anthrax bacilli, Pasteur was also exercising a form of moral authority: *you should do it this way*. In these cases the justification for his moral authority was his record of scientific accomplishment in the area in question. In addition, however, he deployed exceptional rhetorical and political skills, not only in delivering his message but also in garnering support for it. Pasteur offers us a matchless example of how an individual scientist can attain and exercise a charismatic scientific authority as well as moral authority.

# T H REE

In every nation, region, or culturally identifiable society, the contemporary roles of science owe a great deal to the particulars of history and culture. For example, the capacity of science to exercise epistemic or moral authority is vastly different in a Muslim nation such as Pakistan or Egypt than in a Western nation such as Germany or England. This book is primarily about the exercise of epistemic and moral authority by science in the United States. It makes sense for me to limit the scope in this way because, in the first place, American science and culture is what I am best able to interpret and analyze. Second, the United States has been the most influential and productive source of science in the world over the past half century. One might argue that science's entanglements with other societal sectors, and the oppositions that it encounters in its attempts to exercise authority, follow a nigh universal pattern across nations and cultures. I do not, however, believe this to be the case. In societies that depart radically from the tradition of Western democracies, authority is most commonly based on some inner commitment, a component of self-identity quite foreign to Western ideas of the self as an autonomous, self-regulating agent operating within a framework of rights and external, social constraints (Seligman 2000). We have already seen examples of conflicts in the United States between religious conservatism and science, and will see more in chapters that lie ahead. Historically, deference to the authority of revealed religion has limited science's capacity to convey authority on many matters. In contemporary India, Israel, and the Muslim countries of the Middle East, the terms of such conflicts are vastly different. They certainly bear study, but they are beyond the scope of my inquiry, which could be seen as a prequel to a more general study of science's authority in other cultures.

This chapter, then, is a brief history of the rise of science in the United States; in this somewhat episodic scan, there will be pauses for brief accounts of periods and events that have been especially influential in determining our present state. The emphasis is on how these events and periods have formed a peculiarly American sense of science's authority.

### ALEXIS DE TOCQUEVILLE

Alexis de Tocqueville came to the United States in 1831 when he was a young man of twenty-five. He and his friend Gustave de Beaumont made the journey ostensibly to study American prisons, a subject of interest to the French interior ministry. Their ambitions, however, far exceeded that limited charge. They believed that understanding the United States, the only nation in the Western world where the people governed, would set them apart from their peers in France and be the making of their careers. The famous result of their journey of more than nine months, covering more than seven thousand miles, was Tocqueville's *Democracy in America*. His extraordinary book, published in two volumes, was the product of eight years' work. It covered every aspect of American life and culture, seen through the eyes of a well-educated young European of noble background. American science received its fair share of Tocqueville's attention and evaluation:

> Those who cultivate science in democratic nations are always fearful of losing their way in utopian speculation. They distrust systems, they enjoy adhering to facts which they themselves study. As they do not easily defer to the reputations of their fellow men, they are never inclined to swear by the authority of an expert. On the contrary, they always concentrate on finding the weak aspects of his theory.
>
> . . . In America, the purely practical aspect of science is studied admirably, and careful attention is devoted to that theoretical area which is closely related to its application. Americans display, in this respect, an attitude which is always sharp, free, original and productive, but hardly anyone in the United States devotes himself to the

essentially theoretical and abstract aspects of human knowledge. (de Tocqueville 1835, 530)

Tocqueville found in the United States a culture vastly different from the one he knew in France, where there were many universities and schools of higher education, and where the sciences were established as recognized disciplines. The École normale supérieure, founded in 1795, was devoted to the preparation of science teachers. The famous chemist Jean Baptiste Dumas held forth at the Sorbonne, conducting research and lecturing to audiences of six hundred or more. Departments of sciences in which faculty carried out research existed in universities throughout the nation. In short, science and scientists were recognized as central in French cultural life. Small wonder then that Tocqueville concluded: "We must acknowledge that in few of the civilized nations of our day have the higher sciences made less progress than in the United States and that in few have great artists, distinguished poets or famous writers been more rarely found" (1835, 524). Tocqueville may have been quick to come to such a harsh evaluation because it fit so nicely with his overwhelming sense of the United States as a place of frenzied activity in search of fortune: "In America each man finds opportunities unknown elsewhere for making or for increasing his fortune. Greed is always in a breathless hurry; the human mind, constantly diverted from the pleasures of imaginative thought and the labors of the intellect, is swayed only by the pursuit of wealth" (1835, 524).

But science in the United States was not in such dire condition as Tocqueville suggests. The framers of the Constitution shared the eighteenth-century Enlightenment's high regard for science (Cassara 1988). Benjamin Rush, David Rittenhouse, Benjamin Franklin, and Thomas Jefferson were intellectual as well as political leaders, and all shared strong interests in science. They supported the idea that free speech applied to science as well as to politics (Goldberg 1994, 28). As the nation grew, there was lively interest in science among the educated in the cities, and salon gatherings devoted to talk of the latest scientific findings were common. Nevertheless, it was true that in the early decades of the nation, not a great deal of science was actually practiced.

Tocqueville was certainly right in his general assessment of Americans as practically minded doers inclined to be impatient with abstract thinking. Even so, by the 1830s there was a nascent scientific establishment: museums and departments of science arose; Louis Agassiz arrived from Switzerland to establish the Museum of Comparative Zoology at Harvard; and

James Dwight Dana, a professor at Yale, and a distinguished authority on corals and volcanoes, was editing the *American Journal of Science and Arts*. Founded in 1819, the journal was important for its communication of scientific findings to educated laypersons, but over time Dana converted it into a professional journal.

A brief look at the careers of two early American scientists helps put in perspective science in the United States. James Dana is widely regarded as the foremost American geologist of the nineteenth century. His career mirrors in many ways the condition of American science in the middle half of the century. He was born in 1813 in Utica, New York, the eldest of four children of a hardware store owner. He learned early in life that he loved collecting rocks, plants, and insects. At Yale University he studied several sciences under the tutelage of his future father-in-law, Benjamin Silliman, founder of the *American Journal of Science and Arts*. He served as an instructor on a U.S. Navy vessel that sailed to the Mediterranean, where he witnessed an eruption of Mount Vesuvius. His account of that eruption, published in the *Journal,* was his first scientific publication. Back at Yale in 1834, Dana developed a system of mineral classification based on chemical and crystallographic considerations, using the university's collection. He published his *System of Mineralogy* when he was twenty-four years old, and it ran to four editions during his lifetime.

In 1838 the navy launched the U.S. Exploring Expedition, under the command of Captain Charles Wilkes. The expedition, consisting of six ships, had the practical goal of charting islands in the Pacific that could be way stations for American clipper ships and whalers. In addition to Dana, the ship carried civilian specialists in botany and vertebrate zoology, as well as other specialties. Dana had wonderful opportunities to make geological observations, and he drew on them for much of his later career. Some of his observations followed in the wake of Charles Darwin, who had preceded him aboard the British ship *Beagle*, but Dana did make many novel observations and proposed many important theories regarding coral atolls and volcanoes. He pursued his scientific career at Yale until he retired to Hawaii at age seventy to complete his scientific work and revise his texts. He died there at the age of eighty-two.

Dana was not the only scientist of note in United States in that period. The greatest American scientist of the time was Joseph Henry, a truly homespun genius, son of a day laborer in Albany, New York. With the help of well-positioned friends he was able to attend the Albany Academy, free of charge. Henry progressed rapidly, and in 1826, at the age of twenty-nine, he was appointed Professor of Mathematics and Natural Philosophy at the Academy. A fascination with terrestrial magnetism led him to experiment with electromagnetism. He was the first to create an electromagnet by winding insulated wire around an iron core. He built a powerful electromagnet for Yale University that could lift 2,300 pounds of iron, a record at the time. His experiments led him to discover both self-inductance and mutual inductance. Michael Faraday was doing similar work in England, and because Henry did not publish his results promptly, Faraday received credit for discovering mutual inductance. Nonetheless, Henry is credited with the discovery of self-inductance, an important property in magnetism.

In spite of his protestations that he was not the graduate of any college, Henry was made a professor at Princeton University. He was a popular teacher and highly regarded research scientist, almost alone among American scientists of that era in achieving international recognition. When the Englishman James Smithson left a large bequest in 1846 for the founding of an institution in the United States "for the increase and diffusion of knowledge among men," the United States Congress established the Smithsonian Institution. With some reluctance Henry accepted the invitation to leave the life of an academic researcher and teacher to lead the new facility. Under his leadership it became an important source of authority in organizing and setting a direction for American science.

The examples of Dana and Henry demonstrate that the United States was not a wasteland for science. It must be admitted, however, that scientific research was not widely supported, nor was it generally thought to be more than an intellectual adornment, rather than a key to social and economic progress. Those who had ambitions for science in the United States saw the need to improve its social standing. The challenge was to create a credentialed professional caste, which was accomplished in various ways. At Harvard, Yale, and Princeton, science courses were added to the curriculum; students were exposed to chemistry, physics, botany, and geology. The American Association for the Advancement of Science (AAAS), founded in 1848 under the leadership of Joseph Henry and others, was from the beginning organized with the aim of promoting the interests of professional scientists as distinct from interested laymen (Kevles 1995). The United States Weather Service was organized under the Army Signal Corps, in part to make weather predictions of assistance to farmers. During the Civil War, the National Academy of Sciences (NAS) was established as a chartered private organization to provide expert advice to the government. But when the NAS recommended that the various western surveys be consolidated into a federal geological survey, there was opposition from representatives of the states and territories. They had little interest in the prospects for scientific advances that could result from more systematic studies. Science was valued at that time only for the economic fruits that it might bear. In the end the U.S. Geological Survey was founded, but it suffered from a dearth of funding, particularly for more abstract studies.

In the post–Civil War decades the founding of new public universities in the settled states was accompanied by provisions of federal support for practical fields such as agriculture and mechanical arts. The Morrill Act, which made grants of land for public universities devoted to agriculture, was especially notable. Science subjects of relevance to the practical arts, such as chemistry and geology, became an accepted part of the college curriculum, but they were not taught as training for professional work in their own right. Science in the United States in the first decades following the Civil War lacked professional standing; for the most part it was considered ancillary to the main business of colleges and universities, which was to prepare its students for work in the larger world. It was not lost on many leaders of science and their patrons, however, that in Europe, particularly in Germany, universities were pulling ahead by emphasizing the sciences. Faculty devoted to teaching *and* research were producing many new scientific results that did have important applications. For example, the German chemical industry was producing new dyestuffs derived from university-based research or produced by graduates of the newly popular Ph.D. programs (Ben-David 1971).

In response, new American institutions modeled on the German universities were founded: notably, Cornell University, inaugurated in 1868, and Johns Hopkins University, which opened its doors in 1876. Andrew D. White, the founding president of Cornell University, wrote a quarter century after its founding:

> It is something over a quarter of a century since I labored with Ezra Cornell in founding the university which bears his honored name.
>
> Our purpose was to establish in the State of New York an institution for advanced instruction and research, in which science, pure and applied, should have an equal place with literature; in which the study of literature, ancient and modern, should be emancipated as much as possible from pedantry; and which should be free from various useless trammels and vicious methods which at that period hampered many, if not most, of the American universities and colleges.

> We had especially determined that the institution should be under the control of no political party and of no single religious sect, and with Mr. Cornell's approval I embodied stringent provisions to this effect in the charter. (1896, introd.)

White's avowedly nonsectarian vision for the new university met vigorous opposition in ecclesiastical circles. Nearly all the colleges and universities at the time were governed by clerics, and religious themes inserted themselves into all subject areas. Faculty were expected to meet tests of religious orientation. Science took a backseat to concerns with basic Christian doctrine. Cornell and its nonsectarian board of trustees represented a departure from moral and religious formation as a central feature of higher education. The university's founding can be seen as an important early challenge to the authority of religion by nonsectarian thought, in which science was an important element. Indeed, White's two-volume *A History of the Warfare of Science with Theology in Christendom* has been a perennial symbol of the conflict between science and religion in the United States. Even those who criticize it today for errors and inaccuracies concede that it was influential in its time (Stark 2003, chap. 2).

Johns Hopkins University was a different and less controversial matter. It is named after the wealthy railroad financier who in 1867 had a group of his associates incorporated as the trustees of a university and a hospital. He endowed each institution with $3.5 million, the largest philanthropic bequests in U.S. history to that time. When he died in 1873 the funds became available for building the university, which opened its doors in 1878. The stated goal of the institution was "the encouragement of research . . . and the advancement of individual scholars, who by their excellence will advance the sciences they pursue and the society where they dwell." The hospital was completed in 1889, and the medical school opened in 1893. The institution's first president, Daniel C. Gilman, was recruited from his position as president of the University of California, Berkeley. He modeled Johns Hopkins on the German universities, emphasizing research for undergraduate as well as graduate students. Even though Johns Hopkins was avowedly nonsectarian, the choice of Thomas Huxley, an outspoken advocate of Darwin's theory of natural selection, to deliver the opening address in 1873 prompted a negative reaction from those who saw it as an affront to those of religious faith.

Inasmuch as there was little Ph.D.-level science education in the United States, many of the faculty who came to Johns Hopkins had received their graduate training in Europe, mainly in Germany. An exception to the rule was Gilman's first faculty recruit, Henry Augustus Rowland, a young man trained as a civil engineer and teaching at Rensselaer Polytechnic Institute. Rowland came so highly recommended that Gilman hired him a year before Johns Hopkins began operations. Although Rowland, whose interests had turned to physics, was not trained in Europe, he took advantage of the year available to him by traveling in the Continent, gathering the components of his new laboratory, and learning from his visits with researchers there. While visiting the laboratory of the famous Hermann Helmholtz in Berlin, he performed a difficult fundamental experiment that attracted considerable attention. His most notable success at Hopkins, which gained him international fame in the physics community, was the development of the Rowland ruling engine, a device for making highly reproducible gratings for dispersing light according to wavelength.

Ira Remsen was recruited from his position at Williams College to become the head of the Department of Chemistry. He had obtained his doctorate at Göttingen under the prominent chemist Rudolph Fittig, and then served for two years as assistant to Fittig when he moved to Tübingen. His exploits as a chemical researcher and editor became legendary. He and a postdoctoral researcher were the codiscoverers of the artificial sweetener saccharin. He founded the *American Chemical Journal* in 1879 and served as its editor for thirty-five years. In 1902 he succeeded Gilman as president of Johns Hopkins.

With Johns Hopkins and Cornell leading the way, the established universities and colleges began to accord more status to science courses and curricula. Daniel Kevles (1995, chap. 2) tells how both private and public universities underwent curricular reforms. Course offerings, particularly in the form of laboratory courses, were broadened, and degrees in science and engineering were created. Much of this change was opposed by those who felt that the traditional values of a liberal education, emphasizing religion, language, and literature, were being subverted. Fears of conflicts between religion and science were widely held, and the reform-minded new presidents had to tread with care.

Among the better educated in American society there was a growing interest in science. At the same time, among the wealthy businesspeople who came to occupy many of the seats in college board meetings, there was a growing discontent with the lack of practical, real-world relevance in the curricula. This period in American higher education is nowhere more clearly revealed than in the career of Charles William Eliot, who in 1869 became president of Harvard University. Eliot was a departure from the past in many respects. He was not a cleric, as his predecessors had been: he was a product of Boston aristocracy, a Harvard alumnus, a Unitarian, and a chemist. Although he had been a faculty member at Harvard, he left in 1863 to spend two years in Europe, where he studied European educational systems. Upon his return he took an appointment as a Professor of Analytical Chemistry at the newly formed Massachusetts Institute of Technology (MIT). Convinced that American higher education was in need of reform, he laid out his vision in a pair of articles published in the *Atlantic Monthly*, a leading journal of opinion. Shortly thereafter he was elected president of Harvard at the age of thirty-five.

During his sojourn in Europe, Eliot had witnessed many benefits to society emanating from universities, such as the valuable contributions to the German chemical industry flowing from academic laboratories. He realized that American universities could be similarly of service to society if they were to redefine their missions. He proposed that professional schools be reformed, that research laboratories be established, that the curriculum be greatly broadened, and that students be permitted to pursue training in specialized fields of knowledge. Under his leadership Harvard became the nation's best-known and wealthiest university. His example was emulated, with local variations, in many other universities, so that by the end of the nineteenth century, American research universities were becoming a force to be reckoned with in the sciences.

Several factors combined to bring science to the attention of the educated-but-nonscientific American public during the last three decades of the nineteenth century. First, science was beginning to be presented to the public in popularized versions. Edward L. Youmans, who edited *Popular Science Monthly*, had established the magazine in 1872 to present the triumphs of science in language that laypeople could understand. Second, Charles Darwin's *The Descent of Man*, published in 1871, triggered wrathful responses from Protestant clerics and theologians, who strongly opposed statements such as, "We thus learn that man is descended from a hairy, tailed quadruped, probably arboreal in its habits, and an inhabitant of the Old World" (Darwin 1871, 911). The revilements delivered with energy and passion from so many pulpits and lecterns served to put science into the realm of everyday conversation.

John Tyndall came to the United States as a lecturer at just this time. He was born in Ireland in poor circumstances, but rose to become one of the most famous scientists of his era (Tyndall Centre 2008). He is best known as a scientist for his research on the scattering of light. (The Tyndall effect is named for what we see when a beam of light is shone through a liquid in which very finely divided particles are suspended. The particles scatter the light, so we are able to see the profile of the beam as it passes through the liquid.) Tyndall's fame during his lifetime rested on his prowess in disseminating a popular understanding of the physical sciences. He not only made science interesting, but he also vested it with authority—the capacity to say important things about the natural world, things that mattered in the conduct of society and everyday life.

Tyndall's credentials as a scientist were very impressive. In Germany he had done important research and had received a Ph.D., a relatively new degree form. He became a lecturer in physics and in time was widely known for his popular presentations. When he delivered the Belfast Address before the British Association of Science in 1874, he wore the mantle of spokesman for science in general. In that address he argued for the superior authority of science over religious or any other non-rationalist explanation of the physical world. As a speaker he radiated a charismatic authority, buttressed by his reputation as an accomplished scientist. His course of lectures in New York was reported on in the *Popular Science Monthly*:

> Prof. Tyndall's course of lectures in New York has met with a success that is commensurate with the reputation of the lecturer, and the interest of the subject which he has selected for popular elucidation. One of the largest halls in the city has been densely crowded throughout the course of six lectures by the most cultivated and intelligent people of New York and the adjacent towns, and he has been listened to with close and absorbing attention throughout.
>
> But one interpretation can be given to this success, and that is the growing interest in matters of science, and the increasing appreciation of ability in its expounders. [S]uch has been the conquest of prejudice and the enlargement of ideas, that Prof. Tyndall's lecture-rooms, in all the cities where he has spoken, have been filled to overflowing with those who are prepared to accept science on its own merits, without mixing up with it questions of theology. (*Popular Science Monthly* 1873, 499)

Thomas H. Huxley was another famous English scientist who came to the United States as a lecturer. Born in the town of Ealing near London in 1825 into a none-too-prosperous family, Huxley was largely self-taught. At age fifteen he won a medical apprenticeship, and eventually received a sufficient medical education to qualify as an assistant surgeon on a naval frigate assigned to chart the waters around Australia and New Guinea. His studies of marine invertebrates, mailed back to England, brought him acceptance into the ranks of English science upon his return. Because he was not well-off, Huxley wrote popular science articles to support himself, and as a result became very well-known.

Huxley was completely captivated by Darwin's *On the Origin of Species*. In 1859 he wrote to Darwin: "I trust you will not allow yourself to be in any way disgusted or annoyed by the considerable abuse & misrepresentation which unless I greatly mistake is in store for you. . . . And as to the curs which will bark and yelp . . . you must recollect that some of your friends at any rate are endowed with an amount of combativeness which (though you have often & justly rebuked it) may stand you in good stead."

Huxley's persistent and vigorous defenses of Darwin against his many critics earned him the nickname "Darwin's Bulldog." When he came to New York in 1876 for a brief series of lectures, many of the clergy declaimed against the notoriety he received. But the spirit of the times, from the perspective of the cultivated classes, was against the religious activists. The *New York Tribune* wrote: "There has been a disposition among some well-meaning but very injudicious clergymen, to protest against the considerations with which Prof. Huxley has been treated, and to denounce him as an enemy to religion. Such men are in themselves the worse foes to religion—as its bigots have been found to be in every age of its progress. Prof. Huxley will receive the attention due to an eminent man of science, whose candor and fearlessness are only equaled by his ability" (*New York Tribune* 1876).

This period of great clamor and disputation in American society over Darwin's theory of evolution is of importance for our analysis of authority. Darwin's story, and that of the geologists, was in direct conflict with a literal interpretation of the Bible. The prevailing scientific models, based on a mass of observations, were that the earth is ancient and had undergone enormous evolution of both its inanimate geological aspects and its manifold life-forms, including many no longer in existence. Only the most tortured literal interpretations of the biblical account could avoid massive inconsistencies. So in this sense, for many imbued with traditional Christian teachings, there was a conflict of authority. The quarreling was intense, and not at all polite; much was at stake in the struggle for authority between science and religion.

## AMERICAN SCIENCE TURNS PROFESSIONAL

The last quarter of the nineteenth century was a critical period for the development of science in the United States. Basic science—inquiry directed toward the acquisition of knowledge and understanding for its own sake—was on the rise. The new private universities were building their faculties, as were major state universities. Distinct scientific disciplines were forming, as evidenced by the establishment of many scientific societies, including: the American Chemical Society, 1876; the American Physical Society (APS), 1899; the American Geological Society, 1889; the American Society of Zoologists, 1889; and the American Physiological Society, 1887. Science was becoming professionalized; academic researchers were laying claim to a kind of bounded intellectual space where pursuit of fundamental questions about the natural world could proceed without the distractions and turmoil of the larger society, which insistently demanded immediate practical application. Professional societies and scientific institutions, such as the AAAS and the NAS, afforded measures of both prestige and security.

But there was reaction to these changes. The partial withdrawal of academic science from the multitudinous demands of a rapidly growing industrial enterprise contributed to the view, already prevalent in some quarters, that science was of little or no value in fostering economic development. The efforts of people such as Henry Rowland to promote the virtues of "pure science" over invention and application reinforced the conviction of others, such as Andrew Carnegie and Thomas Edison, that higher education in the sciences was useless as preparation for the practical demands of industry (Kevles 1995, 45–47).

As the century wore to a close, science's roles in society continued as a perennially hot topic. On the one hand there was growing support for the idea of scientific research as an unfettered, disinterested search for new knowledge. The University of Chicago, opened in 1891 opposite the grounds of the planned World's Columbian Exposition, was a highly visible expression of this sentiment. Generously funded with an initial endowment from John D. Rockefeller, and supported by the business community of Chicago, the new institution was committed to the primary task of fundamental investigations.

Meanwhile, economy-minded members of Congress were threatening to cut expenditures for what they deemed superfluous science in various federal agencies (Kevles 1995, 62–65). It was a difficult time to obtain budgetary support for scientific research that did not promise immediate application. At the peak of the Panic of 1893, as much as 18 percent of the workforce was unemployed; it did not seem appropriate to spend public monies on activities that could not promise quick contributions to the economy. The country's financial outlook began to improve in 1896, however, and by the turn of the century federal budgets for science-related activities sharply increased. In 1901 Congress approved a bill to establish the National Bureau of Standards, which was to be devoted to whatever research might be needed to form the basis of establishing standards. Because it opened the door to much basic research, it can be counted as the first government agency devoted to scientific work. Its establishment was a vote of confidence in the capacity of science to pronounce authoritatively in matters of broad public interest.

In 1902, in a remarkable turnabout from his earlier skepticism regarding fundamental research, Andrew Carnegie bestowed on the U.S. government the enormous sum of ten million dollars to establish the Carnegie Institution of Washington, meant to "encourage investigation, research and discovery" (Kevles 1995, 69). Meanwhile science departments in the major state universities were rapidly expanding. Increasingly it became a requirement for faculty members to hold a Ph.D., and research and scholarship were expected, in addition to teaching and other academic activities. American science was well on its way to taking its place on the world stage. Even as it did so, however, its roles in society continued to be questioned and discussed.

Many were unsympathetic to portrayals of science as a disinterested search for truth about the natural world, and a virtuous activity that enhanced culture by satisfying innate human curiosity. They saw the promotion of basic, curiosity-driven research as a movement to create a privileged intellectual authority. For example, the pragmatist economist and philosopher Thorsten Veblen, writing in 1908, contrasted basic science with a more pragmatic, or practically oriented, view of the world: "Insofar as touches the aims and the animus of scientific inquiry, as seen from the point of view of the scientist, it is a wholly fortuitous and insubstantial coincidence that much of the knowledge gained under machine-made canons of research can be turned to practical account." In Veblen's view, the employment of scientific knowledge for useful ends occurs without any necessary involvement on the part of scientists: "These useful purposes lie outside the scientist's interest. It is not that he aims, or can aim, at technological improvements. His inquiry is as 'idle' as that of the Pueblo myth-maker" (Veblen 1919, 16). Veblen goes on to contrast the spirit and practice of science with the large body of nonscientific wisdom and practice held in society. He asks rhetorically: "How far is the scientific quest of matter-of-fact knowledge consonant with the inherited intellectual aptitudes and propensities of the normal man? and, What foothold has science in the modern culture?" (21).

## THE MAKINGS OF A SCIENCE ESTABLISHMENT

The new century brought with it great opportunities for science, as well as new forms of challenge to its rise in society. The Progressive reform movement was in its heyday, which meant that science found many places in government where it could exercise authority—for example, in regulatory agencies connected with food, and with weights and measures. The Bureau of Chemistry within the Department of Agriculture was originally charged with evaluating the safety and purity of food and related products, but it gradually metamorphosed into a regulatory agency after passage of the Pure Food and Drug Act of 1906. The work of such agencies made science more visible both to ordinary citizens and to lawmakers. The authority of government scientists derived from their evident contributions to a safer and more efficient society. As science became caught up in the give-and-take of democratic politics, however, the constraints of serving competing interests or of bowing to the pressures of politics grew increasingly onerous. Further, there was little opportunity for basic research in most government laboratories; the task was largely one of applying existing knowledge to specific, practical problems. As science grew to become an important sector of national life, scientists experienced increasingly the opposing tugs of power, "purity," and autonomy.

The image of science in the eyes of ordinary Americans rose immensely as a result of scientists' contributions to the nation's efforts in World War I. George Ellery Hale and other National Academy leaders approached president Woodrow Wilson in 1916 with the proposal to form a National Research Council (NRC), which would stand ready to support pure and applied research directed toward achieving national security. Hale, the son of a wealthy Chicago businessman, had at an early age built an outstanding reputation as an astronomer. He was also an energetic and persuasive spokesperson for science. He had, for example, persuaded Andrew Carnegie to fund the Mount Wilson Observatory. President Wilson in due course did request that the National Academy form the Council, which quickly garnered widespread support in academic, governmental, and industrial circles (Kevles 1995, 111–12). When the United States entered the war, Hale was already pushing an ambitious agenda for the Council on many fronts. German submarines were a great threat to the war effort, so physicists were called on to work out new and more effective means of submarine detection. Chemists helped to develop new explosives, and formulated protections against poison gases. Researchers in the biological and medical sciences made important contributions to the war effort in areas of nutrition, sanitation, and the treatment of war wounds, including gas poisoning. George Daniels summarized matters this way:

> Such concrete evidences of the potency of science increased the authority—both within the scientific community and without—of anyone who spoke in the name of science. . . . The Chairman of the National Research Council, in drawing up a balance sheet of the effects of the war on science, mentioned as gains the better public appreciation of the usefulness of science, increased governmental appropriations for research in several countries, the increased appreciation of the value of science in industry which had led to a multiplication of industrial laboratories, and the recognition by educational institutions that science should have a larger role in the curriculum. (1971, 293)

The sciences had indeed risen substantially in visibility and esteem. In 1922 the NAS established a new location, occupying a prominent place on the National Mall in Washington, D.C. Hale, then honorary chairman of the NRC, described the splendid new facility "as a means of keeping the public in touch with the progress of science and to demonstrate the importance of research" (1922, 515). But Hale was looking beyond the present: he wanted American science to compare with the best the world had to offer. And he was mindful of the continuing imperative to justify basic science as a counter to the views of highly visible skeptics such as Thomas Edison. In reflecting on the prospective visits of budding young scientists to the exhibition rooms of the new Academy building, he wrote:

> Hitherto the United States has produced few great physicists. Is it not probable that some of these boys will be led to recognize the fundamental importance of science and to see with Carty, Whitney, and other leaders of industry that the greatest advances arise, not merely

from the direct attempt of the inventor to solve some special problem, but even more truly from the pioneer work of the scientific
investigator, who discovers the phenomena and formulates the laws
that underlie and render possible both invention and industry? "You
can't have applied science unless you have science to apply," and
the industrial research laboratories now move so closely in the wake
of the physicist and chemist that the scientific discovery of to-day
becomes the working device of to-morrow. (518)

The tirelessly promotional Hale had the ambition to build a new institution for scientific research in Southern California, to complement the
staff of the Mount Wilson Observatory. To that end, he reshaped Thorpe
College, a small technical college in Pasadena, by raising a substantial
endowment from wealthy Southern Californians. Hale recruited the eminent chemist Arthur A. Noyes to head the new institution's chemistry
division, and Robert A. Millikan to head the physics program. Millikan
moved from the University of Chicago in 1921 to become the chief executive officer of the new institution, to be called the California Institute of
Technology.

Millikan had an impeccable scientific reputation, solidified by his receipt
of the Nobel Prize in Physics in 1923. He was prominent during the 1920s
for his attempts to reconcile the advances of the natural sciences with non-
scientific sensibilities regarding values and religious beliefs. Science had
emerged in the postwar period as a powerful force for improvement of
the quality of life through technological applications of basic discoveries.
At the same time, it created deep-seated feelings of unease among many.
As the reigning theories of physics and chemistry grew ever more sophisticated and mathematically grounded, the prospects for general popular
understanding of what scientists were producing declined. Raised in a
clergy household in Iowa, Millikan never lost his conviction that science
and religion could and should coexist. In lectures and in writing he campaigned tirelessly for a reconciliation of the two (Kevles 1995, 177–79). For
example, he formulated and secured signatures from both religious and
scientific leaders for a statement on the relations of science and religion,
published in the journal *Science* in 1923, which read, in part: "The purpose
of science is to develop, without prejudice or preconceptions of any kind, a
knowledge of the facts, the laws and processes of nature. The even more
important task of religion, on the other hand, is to develop the consciences,
the ideals and the aspirations of mankind" (Millikan 1923, 630).

Those with religious convictions were not the only ones to challenge the increased authority of science during the 1920s; humanists felt that science was not only indifferent to many important values, but positively destructive of them. The coupling of science with industry certainly helped to advance the cause of science in government, but it had the adverse effect of associating science with some of the ills of a modern industrial society: a loss of individuality and craftsmanship in the workplace; environmental degradation; and the clamor and ugliness of urban life. During the 1930s there were those who wanted to blame science for a failure to solve the nation's economic ills during the Depression. But there were notable voices on the other side. The popular philosopher and educator John Dewey, writing during the 1930s, believed that science had the potential to liberate society from many of its debilities:

> The general adoption of the scientific attitude in human affairs would mean nothing less than a revolutionary change in morals, religion, politics and industry. The fact that we have limited its use so largely to technical matters is not a reproach to science, but to the human beings who use it for private ends and who strive to defeat its social application for fear of destructive effects upon their power and profit. A vision of a day in which the natural sciences and the technologies that flow from them are used as servants of a humane life constitutes the imagination that is relevant to our own time. A humanism that flees from science as an enemy denies the means by which a liberal humanism might become a reality. (1939, 459)

In spite of various criticisms there was steady progress in advancing the status of science, both at home and internationally. In the period from 1920 through 1940, five American physicists and three American chemists won or shared in Nobel Prizes. All but one, Peter Debye, were native-born Americans. In some large sense, science had indeed acquired authority; it had become an irreplaceable facet of American life.

### WORLD WAR II: A WATERSHED FOR AMERICAN SCIENCE

As the 1930s drew to a close and the war in Europe got under way, the debate on the role of the United States in the world was in full sway. When Henry Luce published his essay "The American Century" in *Life* magazine in February 1941, he called for Americans to take on the challenge of world leadership by appealing to a characteristically American sentiment: "This nation cannot truly endure unless there courses through its veins from Maine to California the blood of purposes and enterprise and high resolve." Luce identified the ascendancy of American science as key to what he saw as the nation's destiny to assume world leadership.

The advent of U.S. involvement in World War II rendered moot all discussions of whether the United States should isolate itself as much as possible from the conflicts in Europe. Scientific discovery and its associated technological implementations were among the most important wartime arenas for action. Even before the United States entered the war, it was evident to scientists in Britain and the United States that the course of the war would depend on technological innovation. It was anticipated that German scientists would make important contributions to the Nazi war machine. The Allied effort to develop an atomic weapon, the most gigantic military development program in human history up to that period, was seen by the scientists involved as a competition between themselves and scientists in Nazi Germany. The international community of nuclear physicists became aware in 1939 of a nuclear fission chain reaction involving heavy elements, notably uranium (Rhodes 1986). The émigré Hungarian physicist Leo Szilard, living in the United States, had realized as early as 1933 that it might be possible, using the fission of heavy atomic nuclei, to create a chain reaction that would release enormous amounts of energy. Now, in 1939, amid World War II, it was clear to Szilard and his close confidantes, notably émigré physicists Eugene Wigner and Edward Teller, that the newly announced fission reaction could form the basis of a bomb, and that German scientists would quickly come to the same conclusion. They felt that somehow they must convey the dreadful potential of a nuclear weapon to the highest authorities. In the end they determined to approach the president of the United States through the good offices of Alexander Sachs, an economist and investment banker who had direct access to Franklin D. Roosevelt. Szilard was on friendly terms with Albert Einstein, and persuaded him to author a letter that Sachs would deliver and explain to the president. The correspondence, crafted jointly between Einstein, Szilard, Wigner, and Teller, but signed only by Einstein, was given by Sachs to the president on October 11, 1939, with an explanation of what nuclear fission could mean as the basis of a weapon and as a new source of energy. Roosevelt ordered that the matter be taken under immediate consideration. What followed was the Manhattan Project, the secret wartime effort that would eventually employ more than one hundred thousand people and result in the production of the bombs that destroyed Hiroshima and Nagasaki, Japan.

There is no more fateful example of the exercise of scientific authority than Einstein's letter to Roosevelt. Szilard, Wigner, and Teller were all physicists of outstanding accomplishments and standing in their fields, but, aside from Szilard, who seemed to know nearly everyone who mattered, they were not known outside the physics community. Einstein, on the other hand, had enormous charismatic authority. He had been recognized as among the greats of modern physics since the early years of the twentieth century. Observations of the bending of starlight on the occasion of the solar eclipse of May 29, 1919, which confirmed the predictions of his general theory of relativity, made him an international celebrity. It did not sit well in the anti-Semitic atmosphere of postwar Germany that the most famous scientist in the world was not only a Jew but also an outspoken advocate of Jewish causes. Einstein saw where Germany was heading, and in 1932 he accepted a position at the newly established Institute for Advanced Study in Princeton. It was to be his home for the remainder of his life.

Once settled in his adopted land, Einstein lived his reputation as an iconic figure, revered and respected by millions who had only the faintest understanding of his scientific accomplishments. Einstein's charismatic authority was probably essential to obtaining an audience with Roosevelt; Alexander Sachs was eager to be the bearer of an important message from the incomparable Einstein. But nuclear physics was not one of Einstein's particular research interests. All the detailed questions that would determine whether a bomb was feasible were not ones that he could speak to with any special expertise. It required the expert authority of the nuclear physicists to move the idea along.

The basic and applied research programs carried out in the United States in pursuit of wartime objectives dwarfed any previous efforts by industry or the government. Scientists of all descriptions were recruited into wartime research. In his book *The Physicists* Daniel Kevles devotes a chapter to what he labels "A Physicists' War" (1995, chap. 20). Certainly the physicists were at the forefront of much wartime research, but scientists and engineers of many disciplinary backgrounds played vital roles in projects crucial to the development of the atomic bomb, radar, synthetic rubber, and the large-scale production of penicillin. The war effort elicited an unprecedented outpouring of invention and technological adaptations. At war's end, the scientific enterprise was a powerful and highly visible element in American society. In important respects, the wartime footing of science really did not end. The standoff with the Soviet Union that quickly followed the end of hostilities meant that there would be continuing dependence on science and technology to maintain military readiness.

THE POSTWAR ERA

World War II had made science a fixture of American life in new and very prominent respects. War fighting had become highly dependent on scientific and technological development. With the onset of the cold war, the development of new weapons and counters to them grew to be a very expensive element of the defense budget. This meant that there would be a continuing demand for highly trained scientists and engineers. Also, the vast amount of research and development undertaken during the war had resulted in a plethora of new technologies that had potential for applications beyond military ends. Examples included applications of atomic energy, particularly the development of nuclear power; jet engines in civilian airlines; microwave technologies in applications such the microwave oven; antibiotics for treatment of a host of illnesses; and instrumentation that could be applied to advance basic research in chemistry and biological sciences. Science and technology, which had made it possible to win the war, unavoidably became a prominent fixture of the postwar world, and the training of scientists and engineers became a national policy concern.

Vannevar Bush, an engineer who at war's end was director of the Office of Scientific Research and Development, offered the single most influential vision of the role of science and engineering in postwar society. Bush graduated in 1917 from MIT with a degree in electrical engineering. Shortly thereafter he became immersed in antisubmarine warfare research. At the end of World War I he joined the faculty at MIT and rose to become dean of engineering in 1932. He had an evident talent for scientific leadership. In 1939 he became president of the Carnegie Institution of Washington, a prestigious position from which to influence the direction of research toward military needs. In June 1940 he prevailed on President Roosevelt to create the National Defense Research Committee (NDRC), with himself as chair. This new committee was given broad powers to coordinate military research efforts across government agencies. In 1941 the NDRC was subsumed into the Office of Scientific Research and Development, a civilian agency funded independently of the military, and Bush served as director. The new agency had broad powers to oversee all wartime research efforts, including the Manhattan Project.

As the war drew to a close, Vannevar Bush was perhaps the most influential and powerful figure in the political world of science and technology. He saw clearly that research in science and technology during the war had been heavily biased toward applications. While this emphasis had served admirably, a new plan for science was needed as a guide for the future. At Bush's urging, Roosevelt commissioned him in November 1944 to prepare a report on how the federal government could promote science in the postwar era. Bush's response—*Science: The Endless Frontier*—was delivered to the new president, Harry S. Truman, in July 1945. It was widely acclaimed at the time and has since remained an influential source of standards for the appropriate roles of basic and applied research under government sponsorship (Bush 1945; Kevles 1990).

During the period between the two world wars, the characteristic inclination of American culture for practical results had made for relatively small outlays by both the government and industry for basic, undirected science. Colleges and universities, where much of the basic research was conducted, had been particularly shortchanged. But the exigencies of wartime had prompted many important applications of basic research that had heretofore gone relatively unappreciated. Bush was convinced that to ensure national security in both military and economic terms, a sound national policy would have to include funding for long-range basic research across a wide range of science and technology. Furthermore, he was convinced that a civilian agency should control decisions regarding policy making and funding related to basic research in colleges and universities. He was elitist in his conviction that research support should go to institutions and investigators on the basis of excellence, regardless of geographical location.

In his carefully crafted rationale for governmental support of science, Bush cited several areas in which it would play a critical role. While military security was seen as crucial, he cited first the need for basic research in medicine and its underlying sciences, all done with the goal of improving human health. In addition, he viewed research in areas far removed from any potential military application as a valuable vehicle for producing a scientifically trained workforce, both for military preparedness and for industrial growth and development. Although he never directly spoke explicitly of "science for science's sake," he was firm in asserting precepts that supported the ideals of autonomy and freedom of inquiry. Describing the new agency that would become the National Science Foundation (NSF), he wrote: "Such an agency should be composed of persons of broad interest and experience, having an understanding of the peculiarities of scientific research and scientific education. It should have stability of funds so that long-range programs may be undertaken. It should recognize that freedom of inquiry must be preserved and should leave internal control of policy, personnel and the method and scope of research to the institutions in which it is carried out" (1945, 9). He went on to further press the case for autonomy:

> The publicly and privately supported colleges, universities, and research institutes are the centers of basic research. They are the well-springs of knowledge and understanding. As long as they are vigorous and healthy and their scientists are free to pursue the truth wherever it may lead, there will be a flow of new scientific knowledge to those who can apply it to practical problems in Government, in industry, or elsewhere.
>
> Scientific progress on a broad front results from the free play of free intellects, working on subjects of their own choice, in the manner dictated by their curiosity for exploration of the unknown. (121)

There were others in Washington with similar ideas on how to support science and technology. Senator Harley M. Kilgore, a New Deal Democrat from West Virginia, had put forth a similar program to Bush's, but—as might be expected for a longtime elected official—he had more egalitarian inclinations with respect to distribution of the funds. In the end, a compromise was reached, and the NSF was created by Congress and approved by President Truman in March 1950. The Foundation was to be governed by a presidentially appointed director, who would formulate policy in conjunction with a National Science Board, made up of presidentially appointed members. Bush was fearful of governmental interference in the affairs of the agency, which would result in the loss of the autonomy he felt was essential to the conduct of scientific research. He lost his fight, however, to make appointment of the director a responsibility of the science board and not the president.

In its early years the NSF's budget was dwarfed by those of defense agencies, especially the Office of Naval Research. Nevertheless, academic scientists generally had considerable autonomy in terms of subject matter and methods employed. The Atomic Energy Commission (AEC), which from its beginnings had a strong bent toward work of military significance, also spent large sums in support of academic research. Even today the NSF is a comparatively small source of funding compared with other federal agencies. The National Institutes of Health (NIH), which early on "captured" research related to the biomedical sciences, is a much larger agency supporting medical sciences, including a great deal of basic science in chemistry, biophysics, and similar areas. The NSF is distinguished by the breadth of its support across science. It may be virtually the sole source sustaining some fields.

There is much evidence that in the immediate postwar period science was valued primarily as a vital element in military preparedness and cold war diplomacy. As noted, most of the available funding came from the Department of Defense or the AEC. During the 1950s, security restrictions were placed on scientists working on certain types of research. Efforts were made by several members of Congress to require security clearances for all recipients of NSF fellowships. Such a requirement was in fact imposed on recipients of AEC fellowships (Kevles 1990, xvi). Later, during the Nixon administration, politics often trumped science. Appointments to science advisory and administrative posts, even the directorship of the NSF, were conditioned on candidates' views on the Vietnam War and other politically sensitive issues of the time.[1] During the Reagan administration, restrictions were placed on communication of scientific results regarding certain subjects deemed critical to national security. Scientists were enjoined from participating in international conferences dealing with sensitive "critical technologies" and from submitting materials in proscribed subject areas for publication. In chapter 7 we will explore several contemporary facets of the complex relationships between government and science.

As American science grew to be a diverse and costly enterprise, nearly every practicing scientist needed external support. In universities and colleges many postwar scientists felt that science could have it both ways: generous support from the federal government—or to a lesser extent from industry and foundations—with virtually no strings attached. Most scientists were only too happy to stay clear of science politics. So long as they were funded to do what they wanted to do, they were happy. For the most part scientists subscribed to an image of science as a community with its own standards of excellence and ethical norms, somewhat withdrawn from

---

1. The same efforts to control science and interfere in its workings were reprised during the administration of George W. Bush; see chapter 7.

the rest of society. The social scientist Robert Merton conceptualized the scientific ethos in terms of four general principles or norms: *universalism*, the notion that a scientific claim is tested in terms of preestablished, impersonal criteria; *communism*, the idea that there is a common ownership of scientific findings; *disinterestedness*, the absence of personal goals, ambitions, and desires in determining conduct as a scientist; and *organized skepticism*, "a temporary suspension of judgment and the detached scrutiny of beliefs in terms of empirical and logical criteria" (1973, 277).

For those who championed the cause of science, the individual strivings of scientists across all the disciplines made for a benign result that could serve as a model for all of human society. Jacob Bronowski, a widely read scientific humanist who narrated the popular bbc television series *The Ascent of Man*, saw in science a virtuous community extending beyond a mere shared methodology:

> I take a different view of science as a method; to me, it enters the human spirit more directly. Therefore I have studied quite another achievement: that of making a human society work. As a set of discoveries and devices, science has mastered nature; but it has been able to do so only because its values, which derive from its method, have formed those who practice it into a living, stable and incorruptible society. Here is a community where everyone has been free to enter, to speak his mind, to be heard and contradicted; and it has outlasted the empires of Louis XIV and the Kaiser. (1956, 67)[2]

For scientists, autonomy meant the freedom to set the rules and norms for practice within their own community. But those who paid for science—whether industry, private institutions, or the state and federal governments—had also purchased the right to look inside the walls and to pass judgment on what science produced, in terms of its social values. Research programs could be terminated for reasons having nothing to do with the scientific merits of the work. Various kinds of restrictions might be imposed—for example, on research subjects, methods of procedure, or dissemination of findings. In short, American science was, and is now, in thrall to many interests outside itself (McGrath 2002). As we examine the ways in which science interacts with other societal forces, such as religion, the courts, and

---

2. Bronowski was British, but he was involved in setting up the Salk Institute for Biological Sciences in California and became a resident fellow there in 1964.

government, we will discover the respects in which those interactions provide both a focus for, and limitation on, science's capacities to exercise authority.

The authority and moral authority of individual scientists, as contrasted with that of science as a social sector, represents yet another dimension of science's influences on American society, particularly during the past half century. Figures such as Albert Einstein, J. Robert Oppenheimer, Edward Teller, Linus Pauling, and Hans Bethe drew on their individual charismatic authorities in attempting to influence public policies during the years immediately following World War II. Their attempts were exercises of both expert and moral authority as they engaged in public debates on issues laden with moral and ethical implications, such as arms reductions and nuclear testing. Their scientific authority in all cases was the product of important and highly visible research accomplishment. But that authority may have been of small account in determining whether these scientists won or lost the ensuing public policy battles.

Postwar policies with respect to nuclear weaponry and the testing of nuclear bombs were prominent arenas in which scientists attempted to exercise influence. Scientists who had worked more or less in consonance with one another during the war to produce the first nuclear weapons parted ways thereafter. Teller and Oppenheimer dueled over whether the hydrogen bomb project should move forward and in what manner. Linus Pauling fought against nuclear testing, while other scientists, such as Teller and Willard Libby, campaigned for its continuance. There were those, such as Teller, who enthusiastically supported the militant cold war ideology, while others, such as Oppenheimer, Bethe, and Pauling, pushed for disarmament. Whatever their stance, those who made an effort to exercise moral authority learned that no matter what science may produce, or however prominent a role an individual may play in that production, in the larger social arena the moral authority of science and individual scientists is in competition with other social forces that may, and often do, prevail (Goodchild 2004; Cassidy 2005; Schweber 2000; Bird and Sherwin 2005).

This brief overlook at the historical development of American science has highlighted the close coupling between the growth and increasing sophistication of science and the explosively rapid expansion of the new nation. Science was valued for its capacities to address the manifold challenges posed by the nation's swift geographical expansion and for its contributions to industrial development. It is surely no coincidence that the pace of invention in the United States in the nineteenth and early twentieth centuries was unrivaled anywhere else. But those who valued science for its potential to contribute to the intellectual life of the nation struggled to overcome the prevailing sentiment that "pure" science had little to offer. Resistance to science's growing epistemic and moral authority came in part from those who thought scientific rationalism inimical to religious and humanist values. Others saw long-term basic scientific research, on which a strong scientific base was predicated, as a waste of resources. In the fullness of time, science has attained a powerful and pervasive presence in American life. Nevertheless, the conflicts and pressures of the past are ingrained in the American culture and continually recur in various guises. In chapter 9 I will have more to say on the evolution of science's relationships with the larger society during the latter half of the twentieth century, particularly as they were shaped by the culture of science itself.

# FOUR

Our concern in this book is mainly with the authority of science in society at large, but authority also operates within the scientific community. In either case, reputation counts for a great deal. The authority with which a scientist makes a case for a controversial position ultimately rests on his standing in the scientific community. The weight of scientific evidence he presents is of course important, but because matters of evidence are often in dispute, the esteem in which a scientist is held is also important, especially when the claims made are novel or contentious. The scientific career of J. Robin Warren offers a beautiful example of how this works. Warren, an Australian pathologist, discovered in the course of routine biopsy work that bacilli were associated with tissues taken from the stomachs of patients suffering from gastritis and peptic ulcer disease. He became convinced that inflammation of the gastric mucosa was caused by the bacteria. This conclusion went against the conventional wisdom that peptic ulcers were caused by excessive gastric acidity. Warren and a clinical colleague, Barry J. Marshall, initiated clinical studies during which they succeeded in cultivating from several biopsies a hitherto-unknown bacterial species (denoted *Helicobacter pylori*). They further showed that eradicating the bacterium from the stomach with antibiotics could heal the peptic ulcers. Warren and Marshall, relatively unknown scientists putting forth arguments that went against conventional wisdom, had considerable difficulty in gaining the attention of the medical establishment. After more than a decade of persistence on their part, however, their discoveries had achieved widespread recognition They shared the 2005 Nobel Prize in Medicine for their discovery of the bacterium *Helicobacter pylori* and its role in gastritis and peptic ulcer disease.

Just as the individual scientist must often struggle to receive attention within science, so it is for scientific institutions—whether a learned society, a research institution, or a national academy of sciences—in their attempts to exercise authority in the world beyond science's confines. Science is often in competition or conflict with other social entities, including government, political groups, nonprofit social action groups, and religious organizations, each with its own goals and agenda. Those on the outside may not see science as significantly different from other interest groups. Thus, the motives of scientists or science for espousing a particular position are typically questioned in the same spirit as for any other advocate. This general skepticism dismays those who believe that science is capable of taking objective stances, existing apart from the give-and-take of contending special interests. Indeed, in their view, only in maintaining detachment, by pursuing a disinterested and objective approach to any issue, can science realize the degree of autonomy essential to the exercise of authority. Thus, it is important to our inquiry to know how perceptions of science by society at large are shaped by the standards science claims to uphold.

Let's begin by examining some foundational aspects of the scientific enterprise: ideas about scientific rationality, realism, and the nature of scientific truth. These matters would seem to be the preoccupations of a small group of academic scholars, but as we have seen in previous chapters, they are part of the history of science in society, and continue to resonate in contemporary life. We then ask about the qualities that make for what Michael Polanyi referred to as the "republic of science"—that is, what it takes for science to have standing as a community. Finally, we look at the tools of persuasion employed by individual scientists or scientific institutions in presenting themselves to the larger society.

### TRUTH, REALISM, AND OTHER IDEAS

One of the perennial preoccupations of many philosophers of science, and of some scientists as well, is the idea of truth. What does it mean to say that something is true? Arguments and counterarguments for and against various definitions of "truth" are a veritable cottage industry in philosophy. A thorough summary of this rich field of study would take us too far afield.

It is worthwhile, however, to suggest at least the range of thoughts about the nature of truth, because we will eventually need to confront what a scientist might mean by claiming that a particular statement is true. Peter Lipton's 2004 Medawar Lecture gives the flavor of recent philosophical discussions related to truth (Lipton 2005).

The classic realist notion of truth, called the correspondence theory, holds that a statement about the physical world is true if and only if there exists a fact or state of affairs in the world that corresponds exactly to that statement. Talking about, say, a chair in plain view, it seems a perfectly simple and obviously correct way to view things: "The chair is on the floor." But science is concerned with much less obviously discernable objects than chairs, and the meanings of both "fact" and "corresponds" can be elusive.

Philosophers of science have held varying opinions on the meaning of truth in science. In one view, we cannot usefully uncouple the idea of "truth" from the means by which we come to know it; truth is integral to our ideas of epistemology, how we come to know the world. This suggestion is rejected by many philosophers of science, but—to put my cards on the table—it seems to me productive to think about the contexts in which we make truth judgments. Truths are not entities with existence apart from agents that claim to know them—that is, "truth" is a human judgment. Consider, for example, the statement, "Stone A is more massive than stone B." This claim for truth is a judgment based on experience, whether it be merely hefting the two stones or weighing them on a balance. If there were no humans (or possibly other rational creatures) in existence to make that judgment, the statement itself would not exist—that "truth" would not exist. Our interactions with the world form the basis of our judgments about truth or falsity. In an important sense, we "make" truth. The nature of that engagement with the world, and the forms that our truth claims about it can take, is yet another arena for philosophical debate.

At the far skeptical end of the spectrum of thought about truth, we have the argument of social construction, which Alvin Goldman describes this way: "There is no such thing as transcendent truth. What we call 'true' is simply what we agree with. So-called truths or facts are merely negotiated beliefs, the products of social construction and fabrication, not 'objective' or 'external' features of the world" (1999, 10). This view of truth, in the extreme form given—and which Goldman vigorously disclaims—largely dismisses a meaningful analysis of how humans interact with the physical world; instead, it deals with social transactions abstracted from the milieu of "science in action." Radical social constructivists aim to deny the validity of traditional thought about what it means to say that something is true; in the process, they also aim to challenge science's traditional claims of access to truth. I will have more to say about social constructivism later in this chapter.

Truth and realism are inextricably connected ideas. Realism, or its opposite, antirealism, are concerned with what does, or does not, exist. These questions lie in the domain of metaphysics, the speculative philosophy concerned with the determination and study of what is real. The entities that might be up for consideration include tables and chairs, numbers, God, moral properties, or electrons and quarks. In the domain of science, issues arise in connection with what we should believe about the many unobservable entities that play roles in scientific theories. The hard-core realist asserts that there exists a mind-independent physical world, and that we humans can come to know that world *as it is*—that is, we can aspire to know both that atoms, molecules, electrons, and quarks exist and what they are in themselves. In this view, theories are *literally* true representations of some aspect of the world. Bas van Fraassen has offered the following formulation of scientific realism: "Science aims to give us, in its theories, a literally true story of what the world is like; and acceptance of a scientific theory involves the belief that it is true" (1980, 8). It has been pointed out many times that theories once held to be irrefutably true have proven in time not to be so. That argument against realism, persuasive to some, fails to convince the hard-core realist. In this respect, the key word in Van Fraassen's formulation is "aims": although a theory in its present form may not be entirely correct, it can be approximately so. Further, as science progresses, theories progressively come closer to the literal truth.

Some alternative views of scientific realism approach the subject by taking account of much that has been learned in recent decades about human cognitive capacities, as reflected in language, concept development, and the coupling between the sensory-motor system and neuronal activity associated with abstract thought (Johnson 2006). In this way of looking at things, which George Lakoff and Mark Johnson have dubbed "embodied realism," our understanding of the physical world is grounded in deeply engrained physical experiences that begin prenatally and are subject to continuous development after birth (1999, 74–93). Of course, we also possess innate capacities for coping with the world, including a capacity for language. For the most part, abstract ideas are understood in terms of core metaphorical concepts that form a largely unconscious set of understandings that we apply to interpretations of experience.[1] Embodied realism essentially dismisses much of the traditional argument regarding reality and definitions of truth as largely irrelevant to an understanding of the nature and limitations of human knowledge. It affirms the idea that there is a mind-independent physical world, but contends that our understanding of it in terms of abstract concepts is expressed metaphorically, drawing on the core physical and social experiences that make up our lives. Thus, there is no general, abstract, truth-bearing language that states how the world is. Rather, scientific observations and concepts are understood largely in metaphorical language. Theories, including mathematical theories, are metaphorical descriptions of what we believe reality to be. We have no other way of talking about and understanding those things that lie outside the scope of direct observations in the macroscopic everyday world in which we live (T. Brown 2003).

Many contemporary philosophers of science advocate a receptive attitude toward scientific realism, while acknowledging the importance of the social dimension. Ian Hacking talks in terms of "representation," which can be thought of as an attempt to create a likeness:

> When science became the orthodoxy of the modern world we were able, for a while, to have the fantasy that there is one truth at which we aim. That is the correct representation of the world. But the seeds of alternative representations were there. . . . In physics and much other interesting conversation we do make representations— pictures in words, if you like. In physics we do this by elaborate systems of modeling, structuring, theorizing, calculating, approximating. These are real, articulated representations of how the world is. The representations of physics are entirely different from simple, non-representational assertions about the location of my typewriter. [My typewriter is on the table.] There is a truth of the matter about the typewriter. In physics there is no final truth of the matter, only a barrage of more or less instructive representations. (1983, 144)

Philip Kitcher argues for what he calls "modest realism." His version contains many of the elements of a traditional viewpoint, such as commitment to an independent reality that exists outside human knowledge,

---

1. For example, the abstract idea of time is expressed in terms such as movement ("In the coming weeks we will see a change"), distance ("That is a long time into the future"), and money ("You're wasting my time"; "I'm running out of time"; "If I could just buy a little time").

and to the idea of "natural kinds" (that entities possess certain essential qualities that enable us to categorize them in some ways that are better than others). Yet he recognizes that scientists work in a social milieu, and that social forces play roles in determining how science gets done. Kitcher socializes the idea of truth by advancing the idea of *significant truth*, which may be epistemic (that is, having to do with knowledge) or practical (that is, having relation to some larger set of concerns). In any event, moral and social values are intrinsic to an assessment of significance. Kitcher asserts that most researchers are not engaged in searching for epistemic significance (some form of truth), or at least not primarily. They have a host of other motivations, deriving among other things from their funding sources, place of work, and the nature of the research problem (Kitcher 1993, 2001).

In *Knowledge in a Social World*, Alvin Goldman describes various approaches to the nature of scientific truth. Rather than trying to decide on a best version, he proposes a more "cautious thesis" about the epistemic authority of science—that scientific practices are superior to "any set of non-scientific practices available to human beings for answering the sorts of questions that science seeks to answer" (1999, 247). He lists six aspects of science that confer veritistic superiority—that is, that empower science to produce more truthful results than any other approach to gaining knowledge of the world (Goldman 1999, 250):

1. An emphasis on precise measurement, controlled test, and observation, coupled with a drive toward continually improved observation.
2. A systematic and sophisticated set of inferential principles (including modern statistics and probability theory) for judging hypotheses on the basis of experimental results.
3. An effective system for marshalling and distributing resources to facilitate scientific investigation and observation.
4. A credit and reward system that provides incentives for scientists to engage in research, particularly in certain directions.
5. A system for dissemination of scientific findings and theories, subject to critical evaluation.
6. The use of domain-specific expertise in making decisions about dissemination, resource allocation, and rewards.

These empirically evident aspects of science as it is organized and practiced provide, in Goldman's view, a solid justification for judging science to be a superior source of truthfulness about nature. The first two have to do with the ways in which scientific practice is conducted, the others with social influences at work within science. The particular scientific "truths" presented to society at large nearly always reflect social influences. Some of these are internal to science, such as availability of resources for pursuit of a given research agenda, ideas within a given subdiscipline about the most promising avenues of research, or competitions among research groups. Others arise from outside—for example, funding decisions, laws regarding permissible research practices, and social attitudes that constrain or strongly influence research in certain areas. This means that some areas of research are pursued vigorously while others lie fallow. It means also that, at any given time, what science has to say about matters of concern to society is a function of its recent history.

Some argue that the nature of science is such that it will eventually arrive at the same truth, even though a variety of factors may determine the timing or the direction in which research is undertaken. Thus, for example, the discovery by James Watson and Francis Crick of the structure of DNA in 1953 was just begging for the discoverer(s) to arrive. If Watson and Crick had not come upon it when they did, someone else would have soon afterward. While this retrospective view may convince us of a certain inevitability of scientific discovery, it must be kept in mind that science's interfaces with society at large are nearly always in the context of the state of scientific knowledge *at present*. For example, in connection with the storage of nuclear wastes under Yucca Mountain, society wants to know what scientists believe *today* to be the case about the longevity of the containment vessels, because critical decisions must be made today. Thus, a combination of external and internal influences imparts an element of contingency, and at times of urgency, to the state of knowledge that science presents to society at any given time.

If it were the case that science, through the employment of all its instrumental and other physical resources, coupled with a set of rational, effective methodologies, could aspire to gain unmediated, true knowledge of the physical world, and if there were a general belief that such was the case, science would be able to speak with unquestioned authority regarding aspects of the physical world on which adequate scientific knowledge was held. Admittedly, we cannot always specify what counts as adequate, but in any case science falls short of such prowess, for reasons to which we have alluded. This does not mean, however, that it is entirely lacking in authority. There is no denying science's deep, largely coherent, and extensive knowledge of the physical world. It is *the* source of our useful understanding of nature, and in turn the wellspring from which technology springs. But for science to exercise authority in society generally, it must have a reputation for acquiring and disseminating reliable knowledge with transparency, free from personal bias, self-interest, or institutional interest. To ascertain the extent to which science can meet such standards, we must look further into the social structure of science, as well as the roles of testimony and trust within and outside science.

## THE SOCIAL STRUCTURE OF SCIENCE

If science is to exercise authority in the larger society, there must be means by which scientists can communicate with one another in order to identify issues, share information, and arrive at consensus positions. But specialization among scientists constitutes a serious impediment to effective communication and community building. It is a truism that no one scientist knows more than a tiny fraction of all that there is to know in science. The great advances made in many areas of scientific research have led to the formation of new disciplines and subdisciplines, each with its own array of journals, meetings, prizes for exceptional work, commercial providers of equipment and supplies, and so on. Increasingly, scientists find it difficult to have meaningful talks about scientific topics with scientists in other fields, even fields that seem closely related. Concerns over this trend have led to increased emphasis on interdisciplinary research, usually performed by multi-investigator groups formed to bring together diverse skills and understandings to attack complex research problems. Such efforts, although important, do not go far toward changing the fact that most scientists are narrowly specialist. It might appear then that science is doomed to be little more than a collection of micro-communities. If science had an impoverished social structure of this kind, lacking a shared, abiding ethos, it could not aim to exercise an authority that rose above that of its members taken individually. How does science avoid the trap of insularity, a pathological tendency toward individualism?

Ludwik Fleck, a Polish-Jewish physician and medical researcher, was a pioneer in the sociology of science. He was one of the first to insist that scientific research, even the production of laboratory experimental results, is a social enterprise.[2] His work was almost completely unnoticed during

---

2. Fleck was born in Lvov, Poland, in 1896. He trained as a physician and later took an active interest in bacteriology. In World War II, Lvov's Poland was swept up into the Soviet empire, and only a few years later was overrun by Nazi Germany. As a Jew, Fleck and his family were sent to the city's ghetto. Fleck was pressed into work producing typhus serum.

his lifetime, but was rediscovered after his death in 1961. Fleck developed the idea of thought-styles: "There exist a certain collective of men possessing a common thought-style. This style develops, and is, at every stage, connected with its history. It creates a certain definite readiness, imparts it by sociological methods to members of the collective, and dictates what and how these members do see" (1986, 72).

Fleck argued for the social nature of scientific cognition:

> No more can cognition be comprehended as a function of two components only, as a relation between the individual subject and the object. Every cognition is a social act, not only when it actually requires cooperation, because it is always based on knowledge and skill handed down from many others. It is social, for during every lasting exchange of thoughts there appear and grow ideas and standards which are not associated with any individual author. A communal mode of thinking develops which binds all participants, and certainly determines every act of cognition. Therefore, cognition must be considered as a function of three components: it is a relation between the individual subject, the certain object and the given community of thinking within which the subject acts; it works only when a certain style of thinking, originating in the given community is used. (154)

Fleck applied his idea of thought-styles to social groups of varying sizes. It applies in varying degrees to large-scale entities such as nations, professions, and science as a whole. It also applies to smaller entities within science, such as a discipline, or in other senses even to individual research groups. While small groups possess varying thought styles, scientists collectively share to some degree in a characteristic scientific thought-style.

The eminent American sociologist of science Robert K. Merton also stressed the social interactions within science that make for a strong sense of community. He identified a set of norms that together constitute an ethos. These norms are to some extent products of history, but they obtain their moral force from science's current practice and structure. Merton

---

He and his family were eventually sent to the Auschwitz concentration camp, and later to Buchenwald. After the war he served for a time as head of the Institute of Microbiology of the School of Medicine of Maria Sklodowska-Curie University in Lublin. His most productive period of scientific research spanned the years 1946–57. His major (and largely overlooked) contributions to the sociology of science, however, were made in the 1930s. In 1956 Fleck and his wife emigrated to Israel, where he died in 1961.

identified four "institutional imperatives" (mentioned in the previous chapter) that constitute the ethos of modern science (1949, chap. 8). The first is *universalism*, the tenet that a scientific result is to be judged independently from personal particulars of the scientist producing it. Fritz Haber's process for the synthesis of ammonia is not invalidated by his complicity with German war making in World War I. The significance of Watson and Crick's proposed structure of DNA does not depend on what we may think of the personality of either scientist.

Second, Merton insisted on the tenet of *communism*. Scientists adhere to the notion of openness, that knowledge is to be publicly shared, and not withheld in the interests of private property rights. Distinguished work is to be honored by awards and other forms of recognition, such as prestigious lectures and the like, and not by financial gain. "Property rights in science are whittled down to a bare minimum by the rationale of the scientific ethic. The scientist's claim to 'his' intellectual 'property' is limited to that of recognition and esteem which, if the institution functions with a modicum of efficiency, is roughly commensurate with the significance of the increments brought to the common fund of knowledge" (610).

Third, Merton identified the importance of *disinterestedness*, a quality that we have previously mentioned. Scientists are expected to keep their personal ambitions and biases in check. The fourth element in Merton's formulation of the scientific ethos is *organized skepticism*: "The suspension of judgment until 'the facts are at hand,' and the detached scrutiny of beliefs in terms of logical and empirical criteria" (614). Merton acknowledges that this characteristic of science frequently brings it into conflict with other attitudes toward a particular topic or set of data. Apart from specific scientific data or discoveries that may bear on a matter, the characteristic scientific attitude of skepticism toward any given belief may be felt as threatening by another source of authority. For example, law places a high value on arriving at a judged conclusion. This process could bypass certain issues about which a scientist could remain uncertain, and therefore make him skeptical of the conclusion reached.

Merton did not believe that scientists are preternaturally disposed toward any of these norms, or that the purported objectivity of science proceeds from the personal qualities of individual scientists. The norms are upheld because of the methodological nature of scientific research, as well as the open communication within science. Steven Shapin contrasts this modern framing of scientific objectivity and reliability with that held by the practitioners of seventeenth-century science:

What underpins scientific truthfulness is said to be an elaborated system of institutional norms, whose internalization guarantees that transgressions will generate psychic pain and whose implementation by the community guarantees that transgressors will be found out and punished. So by the middle of the twentieth century it appears that the causal link posited by gentlemanly culture between truth-telling and free action had been turned upside down. Objective knowledge is not now thought to be underwritten by the participation of "gentlemen, free and unconfin'd," but by institutions which most vigilantly constrain the free actions of their members. (1994, 413)

Merton saw science as exceptional in comparison with other professions in being virtually free of fraud.[3] He did not, however, attribute this to an unusual degree of moral integrity on the part of scientists: "Involving as it does the verifiability of results, scientific research is under the exacting scrutiny of fellow-experts. Otherwise put—and doubtless the observation can be interpreted as *lese majesty*—the activities of scientists are subject to rigorous policing, to a degree perhaps unparalleled in any other field of activity. The demand for disinterestedness has a firm basis in the public and testable character of science and this circumstance, it may be supposed, has contributed to the integrity of men of science" (613).

Merton's depiction of science as a community of like-minded practitioners, bolstered by well-established and generally observed norms, minimizes the role of the individual scientist in creating scientific authority. The credibility of the scientist rests to a large extent on her acceptable credentials and institutional affiliations. The scientist's claims in turn are justified on the grounds that the scientific establishments with which she is affiliated—the institution that employs her and provides the physical context of her research, the agencies that fund her work, the journals that publish her results, the societies in which she has membership and accreditation—all in some sense exert controls that constrain her personal passions and interests in producing a scientific outcome consistent with the four norms that constitute the scientific ethos.

In *Science as Social Knowledge* (1990), Helen Longino is concerned with social and cultural influences on the content and structuring of scientific knowledge. She maintains that scientific inquiry must be understood as a

---

3. In this respect his views have been outdated by developments that have substantially changed the social contract between science and society. More on this in later chapters.

social rather than an individual process. In her model of scientific episte-mology, a scientific belief is justified to the extent that it results from prac-tices sanctioned by the scientist's community, and not simply in some independent, objective sense. She emphasizes the importance of criticism, the responsiveness of beliefs to the outcomes of critical discussion, and a widely shared intellectual authority. In her later work, *The Fate of Knowledge* (2002), Longino addresses what she sees as a false dichotomy between the presumptively rational processes that govern the production of scientific knowledge on the one hand, and the social forces that operate within sci-ence and in its relationships to the rest of society on the other.

These sociological models emphasize the pervasive influence of the in-stitutions of science in forming a scientific ethos and enforcing codes of practice that together constitute the basis of much scientific authority. The challenge for such models is to move from the authority of the individual scientist, which will be highly variable and limited in scope, to a more gen-eral scientific authority. Writing in the post–World War II period, Michael Polanyi attempted just this leap. His intriguing essay "The Republic of Science" anticipates much that has been written since. Polanyi describes science as a "self-coordination of independent initiatives [which] leads to a joint result which is unpremeditated by any of those who bring it about" (1962; see also Polanyi 1945b). In Polanyi's model, the scientist chooses a research area by assessing the depth of the problem and the importance of its prospective solution, primarily by the standards of scientific merit accepted by the scientific community. Thus there is an element of con-formity in the scientist's choice of research, but that is mitigated by the element of originality that attends the choice and the ways in which the research is pursued. Discoveries of the greatest ingenuity are often those that break with accepted communal beliefs. It is this balance, between the guiding role of professional standards and challenges to them, that imparts an authority to science:

> This dual function of professional standards in science is but the logical outcome of the belief that scientific truth is an aspect of reality and that the orthodoxy of science is taught as a guide that should enable the novice eventually to make his own contact with this reality. The authority of scientific standards is thus exercised for the very purpose of providing those guided by it with independent grounds for opposing it. The capacity to renew itself by evoking and assimilating opposition to itself appears to be logically inherent in

the sources of the authority wielded by scientific orthodoxy. (Polanyi 1962, 59)

To move from the individual scientist to a more general source of authority, Polanyi envisions science as a network of individuals with over-lapping competences. Each scientist is generally knowledgeable about a sufficiently broad range of science to make informed judgments about work not greatly different from his own. Each who is a member of a par-ticular small group of specialists, possessed of a particular thought-style, to use Fleck's terminology, will also be associated with similar groups. In this way the whole of science is covered by networks and chains of groups: "Through these overlapping neighborhoods uniform standards of scien-tific merit will prevail over the entire range of science, all the way from astronomy to medicine. This network is the seat of scientific opinion. Sci-entific opinion is an opinion not held by any single human mind, but one which, split into thousands of fragments, is held by a multitude of individ-uals, each of whom endorses the other's opinion at second hand, by rely-ing upon the consensual chains which link him to all the others through a sequence of overlapping neighborhoods" (59–60). In a related vein, Philip Kitcher has elaborated on the role of authority within the scientific com-munity as an important element in determining scientific practice (1992, 244–71; 1993, chap. 8). Scientists gauge the authority of other scientists by deciding whether to trust their results, making judgments on the cred-ibility of novel claims, and other ways that permeate the everyday workings of the scientific community.

How then does the scientific community somehow export the author-ity that is the product of its communitarian culture? Polanyi writes of the authority of *scientific opinion.* He sees it as unevenly distributed among scientists: the more distinguished members of the profession dominate over the less well recognized. But it nevertheless is an authority that is mutual, in the sense that "it is established *between* scientists, not above them. Scientists exercise their authority over each other. Admittedly the body of scientists, as a whole, does uphold the authority of science over the lay public" (1962, 60).

Polanyi's claims for the authority of science, with its elitist overtones, could easily be dismissed as out-of-date. For one thing, the notion that sci-ence speaks univocally on matters of scientific opinion does not hold up well upon close inspection. Of course there are broadly general perspec-tives, such as an emphasis on curiosity-driven research, some precepts of methodology, and the like; but in science's interfaces with the larger society there is nothing like unanimity of outlook. Nor can it be said that uniform standards of scientific merit prevail over the entire range of science. Nonetheless, as John Ziman (2000) notes, the essential features of Polanyi's vision of science remain, despite the inroads of political and market forces. Polanyi and Merton captured important aspects of how the efforts and achievements of individuals contribute to forming an institutional presence for science. Writing in the aftermath of World War II, himself a victim of Nazism, Polanyi was keenly interested in preserving the autonomy of science. He saw the need to present a convincing model for the internal authority of science, expressed as "scientific opinion," to support his claims for scientific autonomy:

> Only a strong and united scientific opinion imposing the intrinsic value of scientific progress on society at large can elicit the support of scientific inquiry by the general public. Only by securing popular respect for its own authority can scientific opinion safeguard the complete independence of mature scientists and the unhindered publicity of their results throughout the world.
>
> . . . Any attempt at guiding scientific research towards a purpose other than its own is an attempt to deflect it from the advancement of science. Emergencies may arise in which all scientists willingly apply their gifts to tasks of public interest. It is conceivable that we may come to abhor the progress of science, and stop all scientific research or at least whole branches of it, as the Soviets stopped research in genetics for 25 years. You can kill or mutilate the advance of science, you cannot shape it. For it can advance only by essentially unpredictable steps, pursuing problems of its own, and the practical benefits of these advances will be incidental and hence doubly unpredictable. (1962, 61–62)

In the model put forth by Polanyi, science's outputs are the products of the coordinated actions of individual scientists, each adjusting his or her aims in light of the state of knowledge within the community. Charles Thorpe (2001) ties Polanyi's insistence on the primacy of the individual scientist's engagement with nature to the writings of Max Weber, produced some thirty years earlier. Weber, in his famous 1918 lecture "Science as a Vocation," asked what could make the profession of scientist—particularly in the academic world—a worthy one in light of the skepticism and disenchantment of post–World War I society generally, and in light of the increasingly bureaucratic nature of professional life. Weber saw that with the increasing complexity of science, and the increasing specialization that follows, the individual could fall prey to the utilitarian spirit of the time and become submerged within the institutions of the state. The only recourse for the scientist who wishes to retain autonomy and integrity is devotion to the search for new knowledge, however small one's likely contribution may be. In so doing, the scientist becomes a model, in society generally, but particularly for students, of how to live a life of service. Steven Shapin (2008) has elaborated on Weber's ideas in light of the state of science in contemporary society. He argues that to downplay the historical importance of individual scientists' moral outlooks has been to overlook a key element in the production of scientific knowledge. Further, the greater complexity and more diverse nature of much contemporary science make even greater demands on the scientist's personal virtues than in the past.

Thus, we have two distinct views of what gives rise to and sustains the authority of science. In one, authority derives from a scientific ethos that emanates from the organizational structure of science, and normative standards peculiar to scientific practices and modes of communication. These may be affected by social influences from outside science, as Longino and others assert, but even these influences are conditioned by the special character of scientific practice. In the other view, scientific authority is vested in the moral characters of its practitioners, who are passionate in their pursuit of "a vision of reality." The scientist is committed to a life of ascetic proportions, the disinterested and "passionate outpouring of oneself into untried forms of existence" (Polanyi 1958, 207). The apprentice in scientific research learns from the master, whose authority is charismatic, deriving from the example of a life lived in science (Polanyi 1945, 1958). These are not incompatible visions; both serve to bolster science's claim that it is a source of reliable, reproducible, and objective knowledge of the physical world, and that the accumulation of such knowledge is progressive.

## SOCIAL CONSTRUCTIVISM

Thomas Kuhn is without doubt the most influential twentieth-century historian of science. In *The Structure of Scientific Revolutions*, published in

1962, he proposed a radically different view of theory change in science. His extraordinarily influential work has given rise to a mountain of literature, as well as new terms of usage, notably the "paradigm" as applied to scientific thought. Here is a much simplified version: Kuhn argued that there is a condition called "normal science" that can obtain in a scientific field, in which a particular theory or set of theories has hegemony. Work in such a field is mostly focused on applying the reigning theoretical understanding to the design of new experiments that extend its reach, or to applications of the theory to new areas. Apprentices in science rise in the ranks by learning how to solve problems in terms of the reigning theoretical and conceptual frameworks. In this situation, normal science proceeds with generally agreed-on theory and methodology, which together constitute a paradigm. But it happens eventually that some results are at variance with the reigning theories, and it proves impossible to accommodate them. After a period of ferment and uncertainty, a new theoretical framework that accommodates the new results is established. But this new theory is generally not expressed in the same "language" as the old one; there is what Kuhn terms an incommensurability between the old and new ways of understanding. Over time, workers in the field abandon the older theory and accept the new one; in doing so they have adopted a new paradigm. Such "revolutions" in science go on at all levels, from rather specialized domains to larger ones.

Kuhn called into question the idea that successive theories represent progress over their predecessors, at least in the traditional sense. The process of theory change cannot be accounted for in intellectual terms only—it is also social in nature: "Persuasion, authority, activism and time all play roles. Whatever scientific progress may be, we must account for it by examining the nature of the scientific group, discovering what it values, what it tolerates and what it disdains. That position is intrinsically sociological" (Kuhn 1970, 238). Kuhn argued that, in fact, a new theory is simply a different way of looking at the world, designed to fit the results in hand. Theories are metaphors, not literal descriptions of how the world is (Kuhn 1979, 533–43). Not much is gained by regarding a new theory as truer than an older one. The new theory simply displaces the old because its language is different.

Kuhn's claims occasioned sometimes-extravagant reactions from philosophers and historians of science. Although his work did not initially receive much favorable notice in the physical and natural sciences, it was a big hit in the behavioral sciences. Before many years had passed, the word "paradigm" had taken on new meanings.[4] During the same period, Paul Feyerabend (1975, 1978), something of a guerrilla philosopher, also achieved some prominence based on his rejection of prevailing ideas regarding scientific rationality and methodology. Many long-held philosophical views on the nature of the scientific enterprise were being called into question by looking more intently at how science actually gets done, and at the trajectories of change in scientific understanding. At the same time, social practices in science and their relationships to scientific authority also came under scrutiny from new schools of thought in the sociology of science.

In the wake of Kuhn's groundbreaking work, new academic subfields of science studies arose, including the sociology of scientific knowledge (SSK). Its practitioners advocated an approach to social studies of science that treats science as an object of study like any other community. Rather than concerning themselves initially with issues such as epistemology, justification, or rationality, they strove to describe and account for scientists' behavior much as an anthropologist might. They focused on methods of persuasion within science, competition, formation of alliances, reputational factors, and the like. They also conducted studies of social practices within research groups, which emphasized how determinations of scientific "facts" come to be arrived at. Whereas earlier sociologists of science might have been content to simply describe scientific activity, the goal of new programs, such as the "strong programme" at Edinburgh University, was to produce a causal account of the content of scientific theories (Barnes and Bloor 1982; B. Barnes 1985; Bloor 1984).

In some social studies of science, the social scientist takes on the character of an ethnographer. The investigators frequently attributed processes of belief formation on the part of the scientists they were studying to social factors (Latour 1987; Latour and Woolgar 1979; Knorr Cetina and Mulkay 1983). Not uncommonly, the outputs of scientific studies were judged to have employed the results of laboratory work or theoretical study to attain aims largely external to the purported business of science (Pickering 1992; Golinski 1998; Longino 2002, chap. 2).

---

4. It might be more accurate to say *set* of meanings. In his postscript to the second edition of *The Structure of Scientific Revolutions* (1962, 174–210), Kuhn wryly noted that one "sympathetic reader" had counted twenty-two distinct uses of the word *paradigm* in the book. Kuhn made a gesture toward clarifying his thoughts on this term, but it didn't really matter much; the word had taken on a life of its own. Not long ago the CEO of a major investment firm was being interviewed in *Forbes* magazine, and was asked how he and his staff decided on which companies to support; he replied, "We sit on paradigm beach and wait for the next big wave" (Sallo 1999).

The challenge presented by constructivist sociology has occasioned vigorous responses by some contemporary scientists and philosophers of science (Kitcher 1992, 1994; Giere 1992; Koertge 1998; Labinger 2006).[5] In any event, is all this Sturm und Drang, sometimes referred to as the "science wars," important for our concern with the authority of science in society? In the world outside academe the arguments among scholars about whether science is objective, or whether it delivers reliable and truthful information about the natural world, go largely unnoticed. Nevertheless, revisionist ideas about science's objectivity and rationality are noteworthy, in part because they reflect broader social shifts in outlook toward authority and trust. Also, they influence how scientists view their profession and their roles in society. As we saw in chapter 3, the archetype of science within the American scientific community—and arguably the Western world in general—has been molded quite actively over the past hundred years or so by key individuals and scientific institutions. But to be credible, that archetype had to be accepted by scientists themselves and reflected in their behaviors.

To reprise the opening theme of the chapter, when it comes to authority, reputation counts for a great deal. Science's reputation in society at large does not rest so much on what scholars in academe think about scientific rationality, realism, and objectivity, as on how science performs in relation to criteria used generally in evaluating social sectors. This means that it depends on the employment of political capital, on economic considerations, on apparent conformity to moral and ethical standards appropriate to scientific work, on how science performs when there are challenges of authority with other societal forces, on familiarity with science, and—running through all this—on effective communication and persuasion.

## TESTIMONY AND TRUST

No one who lives in contemporary society can escape the necessity of trusting others. In the course of our daily lives we repeatedly count on the reliability and truthfulness of others—for example, for information such as the time and temperature, an airline schedule, the tally at the grocery store, or a set of directions obtained online. We also generally trust others to be

5. Labinger deftly points to the inconsistencies in some scientists' criticisms of science studies.

honest when we purchase items, talk to a surgeon about the efficacy of a medical procedure, or receive the advice of a mechanic regarding the need for auto maintenance. When we are dubious about someone's truthfulness, we generally ask more questions and do more investigation to satisfy ourselves about the person's honesty or reliability.

The eighteenth-century Scottish philosopher David Hume is famous for his skepticism regarding our ability to attain truth through the powers of reasoning. In *An Inquiry Concerning Human Understanding* (1748), he claimed an important role for testimony: "We may observe, that there is no species of reasoning more common, more useful, and even necessary to human life, than that which is derived from the testimony of men, and the reports of eye-witnesses and spectators. . . . It will be sufficient to observe that our assurance . . . is derived from no other principle than our observation of the veracity of human testimony, and of the usual conformity of facts to the reports of witnesses" (170). Hume seems to be saying that it is all right to trust testimony when there is empirical justification for doing so. But there are limits; in considering the question of miracles, which he defined as transgressions of a law of nature through divine intervention, he asked how we should react when persons report that they have witnessed a miracle. The credibility of such reports, Hume says, must rest on several criteria, including the reliability of character and the authority of the person. He goes on to enumerate in detail the desiderata for belief that a miracle had occurred, and concludes that no credible evidence for miracles exists: "Upon the whole, then, it appears, that no testimony for any kind of miracle has ever amounted to a probability, much less a proof" (183). In short, when it comes to reports of miracles, people are not to be trusted.

Hume's fellow Scottish philosopher Thomas Reid thought that Hume was generally more skeptical than he needed to be; Reid felt that our tendency to trust the testimony of others, even when there is no prior experiential evidence pointing toward reliability, is a divinely implanted principle of human nature: "It is evident, then, that in the matter of testimony, the balance of human judgment is by nature inclined to the side of belief; and turns to that side of itself, when there is nothing put to the opposite scale. If it was not so, no proposition that is uttered in discourse would be believed, until it was examined and tried by reason; and most men would be unable to find reasons for believing the thousandth part of what is told them. Such distrust and incredulity would deprive us of the greatest benefits of society" (1764, 240–41). What is true of society at large in the matter of trusting the testimony of others holds also for the scientific community. The internal social structure of science is powerfully productive of attitudes and behaviors of trust and testimony, which together go a long way toward constituting its authority. C. A. J. Coady, in his influential book *Testimony*, has this to say: "Any given scientist, even the most authoritative, will argue from, presuppose, and take for granted numerous observations and experiments that he has not performed for himself. That this is so is obscured by the elements of individualist ideology built into our image of science, the scientist being pictured as utterly self-reliant and self-sufficient, and by the way in which writers tend to refer to 'established' observational and experimental fact, as though they themselves had done the observing and experimenting. There is nothing pernicious in this, but it can contribute to a misleading image of science" (1992, 9). There is a certain irony in the notion that trust and testimony are vital elements in the effective workings of science. The seventeenth-century scientists we discussed in chapter 2 asserted the precedence of individual experience and reasoning over the authority of received reports. As pioneers in the use of experiments to study the natural order, they were understandably skeptical of much idle speculation and traditional beliefs, which were ill founded for lack of firsthand observation. The motto of the Royal Society— *Nullius in verba* (On no man's word)—nicely captured their dubiety regarding the value of unevaluated testimony. Nonetheless, as Shapin (1994) has described, such individualistic views necessarily gave way to the demands of effective sharing of scientific knowledge.

As science took on the character of a real social entity, it was understood that testimony would be necessary to the accumulation of practical knowledge and the establishment of consensus on many matters of scientific practice. Indeed, the founding in the seventeenth century of the Royal Society of London and other scientific societies was predicated on just these ideas. The trustworthiness of anyone who would be a member of the community became a salient issue. In this respect science was, and is, no different from other domains of knowledge production: "A hallmark of human culture . . . is to enhance the social fund of knowledge by sharing discovered facts with one another. Communication is an efficient mode of increasing knowledge because information transfer is typically easier, quicker and less costly than fresh discovery. . . . Since not every member of a community observes each fact other members observe, there is room for veritistic improvement through communication. Not every discoverer, of course, chooses to disseminate her newly won knowledge to others, nor is every report or item of testimony a sincere and veridical report" (Goldman 1999, 103). Goldman recognizes the enhanced epistemological efficiency that results from communication, but at the same time he acknowledges that any system of shared communication is subject to lapses or abuse by individuals. How can science retain its credibility in the face of shoddy work or ethical failings on the parts of certain of its practitioners?

A key element in science's claims to be a truthful and reliable source of knowledge lies in its networked nature. While an individual scientist is truly expert about only a narrow field, she is at the same time knowledgeable in more general senses about a much larger body of scientific knowledge. Furthermore, she will know others who are expert in many areas other than her own. She has recourse to those individuals when she has questions about a new report, about the feasibility of a projected new experiment, and so on. In going about her work she places trust in the reports of others about observations on related systems, the efficacy of a procedure with which she is not familiar, and the methods to be used in interpreting instrumental outputs. She may not know any of those whose writings she relies on, other than possibly by reputation. This all works, in part, because scientists share a great deal of tacit knowledge relating to methods, evaluation of observations, and other aspects of scientific practice that are common to science generally (Polanyi 1958, pt. 2). Scientific networking leads to what Philip Kitcher (1993) describes as a "consensus practice." In principle it could be consciously optimized, but in the real world it simply runs, though not necessarily optimally, because that is the way science has for a long time.

So there is an apparent dichotomy: the scientist must be at once skeptical and at the same time trusting. A large fraction of all the experimental results or theoretical computations reported in the literature are taken at face value and never repeated. Sometimes this happens because the results are not very interesting, nor challenging to widely shared beliefs. In a handbook of chemistry and physics, I can find tables of the physical properties of thousands of pure chemical substances. These include melting points, boiling points, densities, colors, and so on. This information is taken on trust; unless I note something that seems quite anomalous, I trust the editor of the compilation to have made good choices from among the available literature sources. On other occasions, as in the case of many high-energy physics experiments, repeating the work would be very costly and time-consuming. On still other occasions, the report may be of a one-off nature, such as observations of a supernova, or meteorological data extending over a particular period of time.

But trust in the work of others is conditional. If a reported result is surprising and flies in the face of expectations based on prior work, those in that particular area of science will evaluate it. Depending on the reputations of those reporting the new results and the seriousness of the implications for future work in the field, there will be quick efforts to reproduce the work, or to design experiments that would further test whatever new hypotheses might have arisen. Of course, confirmation or extension is not always done in a cordial spirit of cooperation. Scientists can be jealous, sensitive of prerogatives, and feel threatened by results that call their own work into question. But whatever motivations move people in cases of controversy, the aim of science generally is to settle disputes one way or another. Even when not all parties to a dispute are ready to concede, the community will generally arrive at a widely held, if not consensual, position. Richard Rorty, a contemporary pragmatist philosopher who persistently denies the claims of science to a special rationality or access to "truth," nevertheless sees the institutional structure of science as the source of its strength:

> There is no reason to praise scientists for being more "objective" or "logical" or "methodical" or "devoted to truth" than other people. But there is plenty of reason to praise the institutions they have developed and within which they work, and to use these as models for the rest of culture. For these institutions give concreteness and detail to the idea of "unforced agreement." . . .
>
> My rejection of traditional notions of rationality can be summed up by saying that the only sense in which science is exemplary is that it is a model of human solidarity. We should think of the institutions and practices which make up various scientific communities as providing suggestions about the way in which the rest of culture might organize itself. (1989, 15)

Science is generally portrayed to the larger society as a largely consensual community with respect to scientific findings themselves. When there is a substantial disagreement between scientists who are well-known outside science, the disagreement may be portrayed as a kind of contest, and be followed like an athletic event. For example, in the race to sequence the human genome, the radical J. Craig Venter claimed to have tools that could do the job more expeditiously than the methods being used in the international Human Genome Project (Davies 2001; Sulston and Ferry 2002; Carmen 2004, chap. 2; Shreeve 2004). The media gave a good deal of coverage to the competing claims of Venter and those who spoke for the Human Genome Project. Eventually, a sort of negotiated joint announcement was made. The message was that the competition had saved time and money, and therefore had been good for science. If there was anger and bitterness in some quarters over how things were portrayed or over who received what credit, the conflicts were kept to a low profile outside the walls of science.

The presentation of science as largely consensual on matters that affect the larger society is but one facet of how science's authority is enhanced by the image it conveys. Another is to be found in the rhetorical style used in presenting science. Most scientists are inculcated with the notion that the scientist must be objective about his work, not influenced by hopes that an experiment will turn out in a particular way, nor moved in the design, conduct, or interpretation of an experiment by ambition, competition, or other emotion. As Karl Popper observed, scientific knowledge is, in the ideal, "knowledge without a knower." Consistent with this belief, reports in the scientific literature are delivered in a style that, insofar as possible, removes the scientist from the picture. Research papers are generally written in the passive voice, and in a stylized manner, varying in only minor ways from one discipline to another. These stylistic devices are themselves tools of persuasion in that they apply a gloss of disinterestedness and objectivity to reported findings and conclusions that may be challenging the correctness of other work or proposing speculative new conclusions. And of course they necessarily omit a great deal that might have been said about experiments that failed, dead ends explored, and other aspects of how the science was actually carried out. In small seminar and conference settings, involving relatively small groups, an informal, candid style of presentation is common, one in which a mutual exploration of unresolved issues between the speaker and audience can occur. Presentations at scientific conferences, in public addresses, and the like, however, have much of the same character of written reports. Often no time is allocated for questions and comments from the audience.

The transformation of science communication from the pages of scientific journals and formal public reports at scientific meetings into materials for consumption in the larger society involves new considerations. What is the audience for the material? What level of sophistication will they have with respect to the underlying scientific issues involved? What are their interests? The audiences involved might include a town council, viewers of a Public Broadcasting Service special, a congressional committee, a group of corporate executives, or a local section of an environmental group. The roles of science in society, from the perspectives of these various interest groups, will be seen in differing ways. We have already seen some of this in chapter 3 in our examination of the history of science in the United States. In the chapters that follow we will be looking at some of the particular venues in which science interacts with society. In some instances, science is making a case for support—perhaps for funding, or for legislation that will permit certain types of research to go forward. In other cases, the authority of science is being exercised to state how things are with respect to some matters of controversy or uncertainty—for example, with respect to climate change, storage of nuclear wastes, or the supply of water in a particular watershed. Whatever the occasion, science relies on its connections with other social actors: "Proponents of science sometimes claim science to be the ideal court of last appeal; in practice, it frequently is an arena of contention, with scientific experts either speaking on different sides of policy issues or declining to take a stand. . . . The authority of science legitimates particular positions and certain groups' powers, and it provides languages and metaphors that shape and limit debate" (Walters 1997, 5).

Science is frequently expropriated to serve interests far outside itself. In a culture overrun with persuasive messages, science is routinely enlisted for a variety of ends and causes: to sell antacid formulations, beauty aids, and a host of other products; to argue for particular public health policies and practices; to justify educational initiatives such as preschool programs; to justify criminal justice practices; and so on. As Ronald Walters has written, "Whatever its ironies and limitations . . . the cultural power of science, as embedded in language and metaphor, helps structure how Americans think and talk about their society" (1997, 7). The range and limits of what Thomas Gieryn (1999) refers to as science's "cultural authority" are the product of continual testing and negotiation vis-à-vis other societal sectors. In the chapters that follow we will examine several specific instances in which science's authority is tested by conflicts with other sectors of society. It will become apparent that the underlying justifications for science's authority outlined in this chapter, and the challenges to them, are actively at work.

## MORAL AUTHORITY

To complete this first part of the book I want to return to a consideration of what we mean by moral authority, and how science or individual scientists can be deemed capable of it. To recapitulate: epistemic, or expert, authority is characterized as the capacity to convincingly relate aspects of the physical world—to tell how the world is. Moral authority, by contrast, is the capacity to instill beliefs as to how the world *should* be—to convincingly argue for particular actions that will presumptively change the way the world is. Most of what has been presented heretofore deals with the grounds for science's expert authority. Even in that telling, however, there have been frequent references to attempts by individuals or one or another institutional scientific entity to exercise moral authority. As we examine the interactions of science with other societal sectors, we will see many more examples. But we need to confront questions of just what it is about the scientific enterprise, if anything, that confers moral authority, as well as the limitations on such a capacity.

In thinking about the nature of moral authority I found it helpful to begin with a rather homely example. Picture Joseph, a man in late middle age, who has had a long and happy marriage with Claire. He has been busy with his career and other matters, and has not been very assiduous in monitoring his health. Claire persuades Joseph that it is time for a complete medical checkup, and he acquiesces. She accompanies him when he meets with the physician, who has examined him and prescribed various tests. Joseph had not seen this particular physician before the checkup.

The doctor tells Joseph that he is in generally good health except for having an elevated level of blood cholesterol. He explains that when left untreated, high cholesterol increases the risk for cardiovascular diseases, particularly heart attack and stroke. He further explains that cholesterol-lowering drugs called statins have been shown in large-scale studies to substantially reduce the risk of heart attacks and stroke. He recommends that Joseph begin taking one of these drugs.

Joseph does not like the idea of taking medications on a regular basis. He has not had to do this before, and for him it somehow smacks of a chronic illness. Furthermore, he is not aware of any adverse symptoms; there just does not seem to be a compelling reason for the expenditure of money and continual worry about taking the pills. In response to Joseph's reservations, the doctor further expounds on the adverse effects of high cholesterol and repeats his recommendation that Joseph begin a regimen of hypertension-reducing medications.

Now Claire speaks up. The husband of one of her close friends had been taking a statin drug to reduce his cholesterol level, and after a time he began to have tingling, burning, and pain in his legs attributable to use of the drug. He began to walk unsteadily and suffered muscular weakness. His doctor then prescribed a different statin drug, but the effects were the same. She tells the doctor that she fears that her husband might become a cripple; she does not think that he should begin taking a statin drug. The doctor replies that while a small percentage of patients may suffer some degree of neuropathy, leading to muscular weakness and other symptoms, the problem occurs rarely and can be dealt with by changing medication or dosage level. In his view the benefits of taking the statins outweigh the risks.

As we analyze this scenario, it is evident that the physician is exercising expert authority in telling Joseph that he has high cholesterol, and in explaining to him the risks associated with that fact. He is not himself the originator of the studies on which his judgment is based; rather, he draws on the work of others in the biomedical literature for the background science, on epidemiological studies relating high cholesterol with increased incidences of stroke and heart attack, and on his own clinical experiences in prescribing the medications for his patients. His medical authority is buttressed by the fact that he is licensed to practice medicine, has graduated from a medical school of some repute, and has been practicing medicine for several years. In other words, his authority as an expert arises from his position within the institutional structure of medical sciences, and on his personal demonstration of having an authoritative grasp of the medical issues.

The physician, however, goes beyond the exercise of expert authority alone by recommending to Joseph that he commence taking the statin drug. He does not simply tell Joseph that such medications exist, that some people take them to good effect, and that he could, if Joseph wished, prescribe one for him. He *urges* Joseph to take the medications, and when the patient shows some reluctance, he reiterates his urgings. In addition, he tries to allay Claire's concerns regarding the medication. In his advice to Joseph and Claire, the doctor is attempting to exercise moral authority. We do not find this the least surprising: giving medical advice is what doctors normally do. We take it as a given that doctors subscribe to some version of the Hippocratic oath, associated with the Greek physician Hippocrates ca. 460–ca. 377 B.C.E.), which provides an ideal of ethics and professional conduct for physicians. In some quarters, even modern versions of the Hippocratic oath are thought to be outdated and not applicable to the complexities of modern medicine. Nevertheless, nearly all medical schools administer a version of the oath to their graduates. There is a culturally embedded expectation that physicians will adhere to a code of conduct in which they are obliged to look out for the welfare of their patients. It is just this expectation that bolsters the doctor's moral authority in advising Joseph.

Claire also exercises moral authority in arguing against Joseph's taking the statin drug. She is not equipped to form an expert judgment on the medical issues involved. What little expert authority she has is borrowed, as it were, from what she knows of her friend's experience. But she possesses moral authority resting in her emotional, social, and financial investments in Joseph. She has made substantial commitments to life with him, and he has a moral obligation to take into account her views in this matter as he weighs his decision. On the other hand, because she lacks significant epistemic authority with respect to the matter at hand, her moral authority is limited.

The example might be extended in many directions. Suppose that Joseph meets his usual group of friends for morning coffee the next day and mentions the visit to the doctor. Andy, the person in this group whom Joseph admires most, tells him that he also faced this decision several months ago and decided to take the medications. His cholesterol level is now down in the safe range, and he has had no adverse effects from the drugs. He urges Joseph to take the doctor's advice. Andy's moral authority in offering advice arises from his empirical experience with the medications, coupled with his concern for Joseph's well-being, a product of their long-term friendship.

To summarize, our example illustrates the two kinds of authority. The doctor exercises expert authority by virtue of his professional qualifications and medical know-how, Claire by borrowing authority from her friend's experience, and Andy by virtue of his experience in taking the medications. These three also exercise moral authority, but as with expert authority, it originates differently in each case. The doctor exercises moral authority by virtue of his expertise and the cultural expectation that as a physician he will act in the best interest of his patients. Claire's moral authority is grounded in her close and long-term relationship with Joseph, as well as their interdependence. Andy exercises moral authority by virtue of his friendship with Joseph. In considering his course of action, Joseph must evaluate their relative moral authorities, and weigh them in light of his own inclinations. While he is doing so, the moral authority of any of the three could suffer as a result of actions they might take. For example, Joseph expects the doctor's advice to be free of self-interest. If he were to read in the local paper that the physician group of which his doctor is a member was taking kickbacks from the pharmaceutical manufacturer of the statin medication being prescribed, the doctor's moral authority might be significantly compromised.

No single document akin to the Hippocratic oath is widely accepted as an ethical guideline for scientists. To exercise moral authority, science depends on society's confidence in the efficacy of science's expert knowledge and reputation for good deeds. To the extent that it is visibly active in societal affairs, has an established record of achievement, and is generally thought to act honestly and in the best interests of society, science can exert moral authority on particular issues. Of the five components of moral authority identified by Gerald Holton (discussed in chapter 1), the two that are most evident to the general public are science's external accountability and its communitarian obligations. These are precisely the two components that require scientists to move beyond the norms suggested by Polanyi, Merton, and others, which pertain mainly to the internal structure of science and justify its authority as a source of reliable expert knowledge. In other words, science's moral authority, *which assumes the validity of its expert authority*, is predicated on what science is seen as doing from beyond the confines of the scientific community. Shapin (1995) has written that science's authority with the public, whether as the authority of the expert or in the exercise of moral authority, rests in large measure on the public's confidence in science as being a reliable source of objective truths about the world. But people outside science who believe that individual scientists are no more virtuous or trustworthy than anyone else may be disinclined to assign a special moral authority to scientists speaking on scientific matters that arise in the public domain. In arriving at an institutional consensus on some matter, science can transcend the more limited authority of individual scientists. Scientists who reflect that consensus view in their communications with the general society derive a moral authority from the association with the scientific establishment that is denied scientists who are outside the mainstream.

Science is a large and important force in society; inevitably, it has become highly institutionalized. Officials of scientific organizations speak for science in such venues as the news media and hearings of congressional committees or regulatory agencies. In addition, many science-related public interest groups advocate positions on various scientific issues. In their attempts to influence legislation or regulation, these officials presume to speak for the scientists who make up their organizations. It is not always the case that individual scientists share the views being expressed. Nevertheless, in purporting to speak for individual scientists, those representing scientific organizations have assumed responsibility for discharging a part of science's external accountability and communitarian obligation.

Some features of the scientific ethos work against formation of a vigorous moral authority for science in society. Scientists engaged in fundamental research often have little interest in practical applications of their work. The strong sense of community that characterizes a collection of scientists working within a fairly narrow domain makes for a strong emphasis on competition and reward structures. In this small society of scientists, the outside world can seem important mainly as a source of funding for research. Participation in activities that pertain to societal affairs outside science requires expenditure of time and effort to learn about the context in which scientific questions arise, as well as the social implications of alternative courses of action that turn on scientific considerations. These are costly diversions from a focus on basic science.

The life of Albert Einstein, the most important scientist of the twentieth century, illustrates the tension for the individual scientist between scientific elitism, with an attendant isolation from outside societal influences, and an active involvement in the social and political life of society. Britta Scheideler (2000) demonstrates in detail how Einstein attempted to sustain two distinct and largely incompatible visions of the place of the scientist in society. In a 1918 speech given on the occasion of Max Planck's sixtieth birthday, Einstein talked about science as an autonomous field, independent of social and political goals. He spoke in quasi-religious terms of the "temple of science," and deplored the cultivation of science for purely utilitarian ends. For Einstein, immersion in science could provide an escape from the world of personal experience. At the same time, he spoke passionately for social justice and democracy and against militarism and blind adherence to any form of prescribed authority. Unlike many of his German scientific contemporaries, Einstein was outspoken in his antipathy to the rise of National Socialism, and he wrote extensively on how democratic principles could be enhanced through the leadership of creative individuals. He struggled to sustain his vision of solidarity within science even as external forces worked against it. When Max Planck yielded to pressure to eject him from the Prussian Academy of Sciences in 1933, Einstein wrote to Planck to say that their friendship remained unchanged. We saw in chapter 3 how Einstein's moral authority was brought to bear to induce president Franklin D. Roosevelt to take seriously the threat of atomic warfare. It was but one of many occasions on which he lent moral authority in the service of causes he deemed important to society at large. When the destructive power of atomic weapons was made manifest at Hiroshima and Nagasaki, Einstein was appalled. Thereafter he lent his name to efforts to contain the spread of nuclear weapons and limit their further development. As recent biographies of Einstein make clear, he was no saint, nor was he especially discerning in matters relating to the interfaces between science and society (Isaacson 2007; Neffe 2007). Nonetheless, his incomparable charismatic reputation imbued his every opinion with a special resonance. In the end, his moral authority depended very little on his epistemic authority as a scientist.

# II

# SCIENCE IN SOCIETY

# FIVE

SCIENCE AND THE COURTS

Because so much of what constitutes the modern world derives from science and technology, it is inevitable that considerations of science and its outcomes are interwoven into nearly every important domain of society: art, law, religion, economics, politics, and government. This chapter is about the ways in which scientific authority is exercised in law, and the special conditions that limit its influence there. Sheila Jasanoff explains why the relationship between science and law is so important:

> Law and science are two of the most important sources of authority for modern governments. Perhaps nowhere is this statement more transparently true than in the United States, where law often completes the work of politics and public affairs, and science as frequently underwrites the rationality of public decisions. As relatively apolitical institutions, law and science are powerful generators of trust. The findings of both are expected to be impartial, disinterested, valid without regard to the immediate context of production, and true insofar as participants in either institution are able to gauge the truth. Social order in democratic nations depends on both institutions living up to this ethos, or at least strenuously attempting to do so. Together, law and science have underwritten a time-honored approach to securing legitimacy in public decisions. (2005, S49)

Despite their similarities, it should not be surprising to find disjunctions between science and law; they make separate and sometimes competing claims of authority, based on strikingly different grounds. We have seen in chapters 1 and 4 that the authority of science rests on its having a rational, objective, and methodical approach to questions about the natural world. Science portrays itself as self-correcting of error, and thus progressive in accumulating knowledge and understanding. Scientists are pictured as disinterested seekers after the truth, bound to fully disclose their findings, even when those findings do not accord with widely held presuppositions. Through processes such as peer review, bad science is filtered out; what is published in reputable scientific journals represents work that meets certain standards. Skepticism of new scientific results or theories is inherent in science; questioning current understandings and moving beyond them is one of the mechanisms by which science progresses. That said, there are large areas of consensus within science regarding the major theoretical posits and what constitutes competent work.

Science's claims to knowledge deal with descriptive matters; they concern the physical world as it is. The courts, on the other hand, concerned with a world defined by laws and legal precedents, are engaged with moral questions. Truth is arrived at by different pathways in science and law, yet there are points of similarity. In both domains, procedural rules screen raw, purported facts for reliability and trustworthiness. We saw in the previous chapter that science is influenced by many social factors that play important roles in determining what stands at any given moment for consensual scientific thought. The processes of rational inquiry—making controlled observations (including data collection), hypothesis formation, experimental testing, and model building—are paramount elements in scientific work. Through peer review, scientific claims are subject to tests such as reproducibility, fit with established science, consistency, clarity of presentation, and so on.

In a somewhat similar way, what counts for truth in law is the outcome of contention over a particular body of information in an adversarial process bounded by rules of evidence. Therefore, law is very much about process. The rules that govern these processes evolve over time, largely through the decisions of appellate judges, who write opinions interpreting and applying judicial precedents to ever-changing factual settings, and—at the highest level—interpretations of the meaning of the Constitution. Judges at the appellate level are also faced with discerning the intent lurking in the sometimes-mysterious language of statutes.

Fact-finding in law occurs through prescribed forms of courtroom discourse. Agents on both sides of a question seek to gain acquiescence in the truth as they present it to the judge and jury. Evidence, testimony, and precedent all form the raw materials for legal decision making. There is no motivation for either side to present matters other than those favorable to its version of the truth, though in criminal cases the prosecution is required to turn over exculpatory evidence and to provide evidence in mitigation at sentencing. In the end, truth is established on the basis of some legal standard, such as "beyond a reasonable doubt" in criminal cases, or "by a preponderance of the evidence" in civil litigations. The materials that furnish the basis for the determination of legal "truth" are restricted to those admitted into the courtroom proceedings.

Scientific authority stands in varying relationships to legal authority. The law may defer to scientific authority in some matters and under some circumstances, but it generally need not do so. The greatly different perspectives of those whose lives are in science and law create many opportunities for talking at cross-purposes (NRC 2002). Many extraordinary legal decisions have flown in the face of compelling scientific testimony. The reception of expert scientific testimony in a legal proceeding is dependent on many factors, some of which would seem to most scientists not relevant to a considered judgment of its scientific worth. Most important, the authority of science is utterly dependent on its findings being admitted into the courtroom. So the first challenge for science in exercising any authority whatever in the legal arena is to get inside the courtroom door.

## ADMISSIBILITY OF SCIENTIFIC EVIDENCE

Scientific evidence and testimony are key elements in many kinds of legal proceedings. Examples include tort litigation related to products alleged to have caused harm; suits in the wake of school board decisions to include creationism or intelligent design in science curricula; challenges to the decisions of federal regulatory agencies on allowed levels of pollutants or regulations of new technologies; and forensic evidence and scientific testimony in various kinds of civil and criminal proceedings. For any testimony or evidence to affect a legal proceeding, it must be admitted. We thus begin our look at science and the courts with a brief look at the history of expert testimony in the American legal system, and a summary of the precedents that currently prevail.

In common law, the opinions of witnesses are not generally admissible in court. That is, a witness is not allowed to opine that Fred is the guy who

did the dastardly deed, or that in all likelihood Agent Orange is a carcino
gen. An exception to this general rule is made for expert testimony. An
early case in English law arose in 1782, when Lord Mansfield allowed an
engineer to give his opinion about the cause of a harbor's filling in with silt
(Golan 2004, chap. 1). He said that "the opinion of scientific men upon
proven facts may be given by men of science within their own science"
(*Folkes* 1782).[1] Mansfield's decision, a kind of legal capstone to the Enlightenment age, opened the door both to expert testimony and an adversarial
process in which the expert could be challenged.

In the second half of the nineteenth century, as science in the United
States began to develop in earnest, and as the Industrial Revolution brought
a host of changes to society, the growing authority of science was reflected
in ever-increasing appearances of expert witnesses in legal proceedings.
Not everyone was happy with this turn of events. It was not clear what
constituted "expert testimony"; expertise, it seemed, could be claimed on
the basis of slim credentials. Further, though experts were nominally serving as disinterested parties, they promoted the cause for which they had
been enlisted. Thus, Lee Friedman wrote:

> The position of an expert on the witness stand, who does not testify
> either to what he has observed or knows as fact but expresses merely
> his opinion as to a situation or on facts which have been established
> by other witnesses, is anomalous in Anglo-Saxon law. . . . While
> the principles on which such evidence is introduced have come to
> be well recognized and while the profession no longer has any reser-
> vations in approving theoretically the use of expert testimony, yet, on
> the other hand, there is a constant complaining and mistrust on the
> part of judges, juries and lawyers of the expert witness. (1910, 247)

In this period, the prevailing standard for admissibility of an expert witness was that there be a need for expert testimony: "The practice of the
courts is to admit the testimony of a class of witnesses who are not supposed to have personal knowledge of any facts or circumstances bearing
upon a pending case, but on the assumption that they are able from their
special training and experience to apply scientific tests and present to
the court and jury the import and value of such evidence as may appear,

---

1. References to legal decisions are collected in the References section under Legal References, and the text citations are in italics.

which laymen could not be expected to comprehend and properly maintain" (Chapin 1880, 365).

Evidence of expertise was generally taken to be success in an occupation or profession that bore some relevance to the matter at hand. Thus, scientists and practicing physicians were exemplary of those who might serve as experts. The field of expertise was supposed to be one that enjoyed broad societal recognition as worthy of respect. Self-identified "experts" claiming phony cancer cures or clairvoyance were excluded. Nonetheless, the vetting of expertise was slipshod, and much questionable medical and scientific advice found its way into the courtroom (Golan 2004; Mnookin 2007). The situation changed in 1923 with the famous case of *Frye v. United States*. The defendant, James Alphonso Frye, was charged with murder. He pleaded not guilty, and offered in his defense the results of an early form of the polygraph test, the systolic blood pressure deception test. The test was new and its validity largely unproven. There was as yet no profession of polygrapher, nor a developed market for their services. Given that there was no established profession on which to base expertise, the traditional standard for admission of expert testimony, what was the judge to do? The circuit court judge in this case issued an opinion containing these key sentences: "Just when a scientific principle or discovery crosses the line between the experimental and demonstrable stages is difficult to define. Somewhere in this twilight zone the evidential force of the principle must be recognized, and while courts will go a long way in admitting expert testimony deduced from a well-recognized scientific principle or discovery, the thing from which the deduction is made must be sufficiently established to have gained general acceptance in the particular field in which it belongs" (*Frye* 1923).

The *Frye* decision, in which the lie detector test offered by the defense was excluded, broke new ground in accommodating legal criteria to the notion of bodies of established knowledge that stood apart from any individual. The expert, however well educated or esteemed in the community at large, also had to be able to assert legitimate ties to established fields of practice. For many years the *Frye* test of "general acceptance" stood pretty much as the standard of admissibility of novel forms of scientific evidence (Golan 2004, 258–59). Along with its salutary effects of excluding quackery, however, it opened the gates for professional groups to maintain relatively uncontested control of what would constitute expertise in a given field. General acceptance within a field is hardly an assurance of worthy testimony if the field to which it applies accommodates faulty science. In addition, with the greatly increased influence of science in societal affairs following World War II, there was a reaction to the growing sway of scientific expertise in legal affairs. Some felt that the courts were delegating legal decisions to scientific experts. Further, the "general acceptance" doctrine placed an unfair burden that could fall on either plaintiffs or defendants by precluding the introduction of expert testimony that could otherwise pass reasonable tests of reliability and validity. For example, it would exclude a promising new forensic technique for which there was not yet an extensive published literature. With the passage of time it became increasingly evident that a new approach to the admissibility of evidence was needed.

The Federal Rules of Evidence, codified by statute in 1975 and modified from time to time, are those that federal judges apply to this day (*Federal Rules of Evidence* 2007). Further, many state courts revised their rules along the same lines. Three of the rules regard scientific evidence. Rule 403 states that otherwise-relevant evidence may be excluded if its probative value is outweighed by the danger that it will be prejudicial, confusing, misleading, or a waste of time. Rule 702 states that testimony is admissible from any qualified scientific expert who possesses "scientific, technical or other specialized knowledge [that] will assist the trier of the fact [e.g., a jury] to understand the evidence or to determine a fact at issue." It states, in part, that "[a] witness qualified as an expert by knowledge, skill, experience, training, or education, may testify thereto in the form of an opinion or otherwise, if (1) the testimony is based upon sufficient facts or data, (2) the testimony is the product of reliable principles and methods, and (3) the witness has applied the principles and methods reliably to the facts of the case." Rule 703 provides that experts may base their opinions on data that might not be admissible as evidence if those data are "reasonably relied upon by experts in the particular field in forming opinions or inferences upon the subject."

## DAUBERT

The Federal Rules of Evidence were not in themselves all that helpful in clearing up a chaotic situation with respect to scientific testimony. Many felt that a great deal of junk science and pseudo-science was flooding the courts, particularly in tort cases (Foster, Bernstein, and Huber 1993; Huber 1993; Angell 1994). In 1993, the United States Supreme Court, in the case of *Daubert v. Merrell Dow Pharmaceuticals Inc.*, issued an opinion that transformed the way in which federal courts (and, in due course, most state courts) deal with the admissibility of scientific testimony.

*Daubert* was a "toxic tort" case. The suit was brought by Daubert on behalf of two minor children of the Daubert family, who suffered from a tragically common set of birth defects. The plaintiffs claimed that their birth defects were the result of their mother having ingested Bendectin, a drug produced by Merrell Dow and widely used in the 1960s and 1970s for alleviation of severe morning sickness. Merrell Dow argued that no competent scientific research showed a causal relationship between Bendectin and birth defects; the company supported their position with expert testimony that included several large epidemiological studies showing no statistically significant relationship between Bendectin ingestion and birth defects in offspring. The plaintiff offered evidence based on animal-cell studies and live-animal studies, chemical-structure correlations between Bendectin and substances known to be teratogenic, and testimony related to an unpublished "meta-analysis" of the existing epidemiological studies. The original trial court dismissed the suit before trial, on the grounds argued by Merrell Dow that the testimony offered by the plaintiff did not meet a general acceptance standard. The trial court decision was upheld on appeal to the United States Court of Appeals for the Ninth Circuit. The plaintiff appealed the case to the Supreme Court, which agreed to hear it. Before the Court, the plaintiff argued that the applicable rules of evidence did not permit judges to screen scientific evidence in the way that they had in this case. Rather, because it was the primary responsibility of the jury, not the judge, to evaluate the evidence (including the science), the case should have gone to trial.

The evidence that the plaintiff wanted to present was in fact quite weak. The research results were not strong enough to support a compelling causal claim, and at best could be regarded as weakly supporting the plaintiff's position. Researchers testifying for the plaintiffs used a meta-analysis to combine the results of several studies, and claimed to find a statistically significant effect, even though none of the studies showed a convincing correlation on its own. The work had not been published in a peer-reviewed journal. Merrell Dow argued that judges have a responsibility to screen all evidence to ensure that it is relevant and trustworthy.

The issue before the Supreme Court was whether the *Frye* criterion of general acceptance was valid. The majority opinion, written by justice Harry Blackmun, repudiated the *Frye* test as incompatible with the Federal Rules of Evidence (*Daubert* 1993). The Federal Rules provided for a liberalization of the standards for admitting evidence as compared with *Frye*. Yes, juries should be arbiters of the value of expert testimony; at the same time, however, the Court stressed that trial court judges must act as "gatekeepers," excluding scientific evidence not based on valid research. Evidence must be not only relevant, but reliable. The Court in effect charged judges with assessing the scientific merit of proffered scientific testimony. The case was sent back to the Court of Appeals for the Ninth Circuit to be resolved in light of the new guidelines laid down by the Supreme Court. The Ninth Court interpreted the new standard rigorously; finding that the plaintiff's claim was not supported by the science put forth, it dismissed the claim.

*Daubert* was seen in the scientific community as an extremely important case, and several big guns of the establishment submitted amicus briefs. One of these came jointly from the NAS and the AAAS. Another was authored by a large group of prominent scientists, including nine Nobel laureates. Still a third came from another group of scientists, physicians, and historians of science, including Stephen Jay Gould and Gerald Holton. These briefs by no means spoke with one voice; in fact, the brief by Gould, Holton, and others was submitted in support of the plaintiffs, whereas the others mentioned were neutral or on the side of the defendant. One major point of disagreement was how much weight to attach to peer review as a measure of reliability of scientific evidence. Most of the briefs emanating from scientists or scientific organizations were alike in arguing for a distinctly scientific authority in determining what constitutes admissible scientific evidence:

> The requirement that an expert opinion must be based in "scientific knowledge" does not simply mean that the subject matter of the testimony must be scientific—e.g., medicine or chemistry or physics. Rather, the limitation imposed by the term "scientific knowledge" requires that the content of the expert's testimony must be developed in accordance with scientific methodology. In particular, the use of the term "knowledge"—rather than "evidence" or "testimony"— underscores the requirement that an expert's testimony be developed in conformance with the rigorous standards of replicable experimentation and reporting that are the hallmarks of the scientific method. Only adherence to accepted scientific methodology can produce "knowledge"—that is, a deduction based upon empirical evidence that can be reproduced, challenged, accepted or built upon by other scientists. (*Daubert*, amicus brief, AMA 1993)

Although the details of science may be remote from common experience, judges can understand the fundamental characteristics that separate good science from bad. Courts should also consider the institutional mechanisms, namely the different forms of peer review, that have been developed to assure that scientists conduct their work in accordance with appropriate scientific criteria. When faced with disputes about expert scientific testimony, judges should make full use of the scientific community's criteria and quality-control mechanisms. To be admissible, scientific evidence should conform to scientific standards and should be based on methods that are generally accepted by the scientific community as valid and reliable. Threshold determinations concerning the admissibility of scientific evidence are necessary to ensure accurate decisions and to avoid unnecessary expenditures of judicial resources on collateral issues. (*Daubert*, amicus brief, AAAS and NAS)

The issue in the present case is quite simply whether the law should impose liability on the basis of testimony that states a conclusion on a scientific subject that is based on methodology and principles that are not generally accepted by other scientists at the time the testimony is proffered. A court can operate on no other principle, for if it admits conclusions based on methodology and principles that are not now accepted by the scientific community, but may in the future come to be accepted, it is operating in the realm of speculation. One cannot today say what may be accepted 50, 10 or even one year from now. Because litigation must terminate, but science intends to continue inquiring forever, the standards of discourse must be quite different. (*Daubert*, amicus brief, N. Bloembergen et al.)

Clearly, in all these statements there is a strong argument that scientific testimony, to be authoritative, must conform to certain standards that operate within science itself. That is, science should be, at least in some general sense, the arbiter of what constitutes valid scientific testimony.

The language of the majority opinion in *Daubert* draws rather eclectically on the twenty-two amicus briefs on both sides of the issue. In general, the Court suggested four factors that should bear on the validity of scientific evidence: (1) It should be based on testable hypotheses, and it should have been tested. (2) The error rate associated with the scientific judgments at issue must be considered; it must not be excessive. (3) The basic research on which the testimony is based should have been published in peer-reviewed journals, or in other ways vetted in the relevant scientific community. (4) The science on which the testimony is based should be generally accepted in the relevant scientific community. (Notice that "general acceptance" crept back into the criteria, but now only as one of several factors to be considered.)

None of these four criteria is determinative in itself, which places a considerable burden on judges. Further muddying the water, the majority opinion borrows quite disparate ideas regarding scientific method and the validity of scientific explanation from two philosophers of science, Karl Popper and Carl Hempel (Haack 2005). Popper argued that the criterion of the scientific status of a theory is its falsifiability, refutability, or testability; Hempel, on the other hand, was concerned with understanding scientific explanations through deductively valid arguments, along with empirical confirmations of theories. Though not trained scientists, judges would be expected to delve into the nuances of philosophically based arguments over the validity and reliability of scientific testimony. Nevertheless, the majority of the Court felt that it was a burden judges could bear. Incidentally, only a few years later, the Supreme Court ruled in *Kumho Tire Company v. Carmichael* that "*Daubert*'s general holdings—setting forth the trial judge's 'gate keeping' obligation—apply not only to testimony based on 'scientific' knowledge, but also to that based on 'technical' and 'other specialized' knowledge." That is, engineers and other experts are to be subject to the same qualification standards (*Kumho* 1999).

The *Daubert* decision, important as it was, merely set the stage. The role of science in the courts is determined by the cumulative case law that rests on *Daubert*. Many law journal papers have gone to great lengths to instruct jurists as to how they might best interpret *Daubert* to most effectively employ scientific expertise. One important example is the paper published shortly after the *Daubert* decision by Bert Black, Francisco Ayala, and Carol Saffran-Brinks (1994). Ayala is a distinguished biologist with a strong interest in the philosophy of science. In addition to his other scientific credentials, he was at the time president of the AAAS. In offering advice at considerable length as to how *Daubert* should be implemented, the paper exhibits a strong implicit assertion of science's authority:

> Though the details of science may be remote from common experience, nonscientists can understand the fundamental characteristics that separate science from pale imitations. Courts can also avail

themselves of the institutional mechanisms science has developed to screen and disseminate scientific information.

. . . Courts must discard the old surrogates for understanding science and develop a new analytical framework based on the criteria and quality-control mechanisms that scientists themselves rely on in assessing each others' work.

. . . We believe lawyers and judges will accomplish the task of understanding science best if they look for the same traits that are important to scientists and if they rely on the same process of review that scientists use. . . . By using these scientific criteria and relying on the review mechanisms of science, the law can take full advantage of scientific expertise while avoiding the pitfalls of false claims posed in scientific garb. (Black, Ayala, and Saffran-Brinks 1994, 720, 721, 753)

In other words, let the science establishment, through its institutional quality control mechanisms, be the model for gatekeeping with respect to the admissibility of scientific evidence and testimony.

So what changes in the treatment of scientific testimony and evidence has *Daubert* wrought? Surveys of judges, juries, and court decisions paint a rather mixed picture, according to a review published by Michael Saks and David Faigman (2005). One effect has been that some fields of expert testimony have attempted to avoid falling under the aegis of *Daubert* by characterizing themselves as something other than scientific. The *Kumho* decision, however, embraces all such areas of expertise. Challenges to the admissibility of expert evidence have increased since *Daubert*. Certain traditionally accepted forms of expert testimony, such as handwriting expertise, have come in for more rigorous evaluation as to admissibility (Mnookin 2001a).

The evidence from studies of legal opinions (Groscup et al. 2002) and interviews with state court judges (Gatowski et al. 2001) suggests that although judges strongly support the gatekeeper role and believe that they are actively fulfilling it, very few of them have a working understanding of the four factors laid out in *Daubert* that they should be taking into consideration. The criterion most heavily relied on is general acceptance. Saks and Faigman conclude from their analysis that "judicial opinions displaying sophisticated application of *Daubert* or other thoughtful focus on the validity of the proffered expertise are few and far between. The major exceptions to this generalization seem to be toxic tort litigation, where judicial sophistication is more evident, and cases involving economic analysis of damages" (2005, 124).

Jurists appear to have been very casual about the validity of forensic science, an area that might have been expected to come in for much more intense scrutiny (Thornton and Peterson 2002; Toobin 2007). One method that might have been probed more extensively for its scientific validity is fingerprint identification. Yet until recently it has not been called into serious question, and it continues to enjoy a high level of confidence from many jurists. Fingerprint identification is a fascinating story in its own right, and offers several cogent lessons for how authority gains ascendancy in the law.

## FINGERPRINT IDENTIFICATION

Jennifer Mnookin (2001b) tells the story of Mark Twain's 1894 novella *Pudd'nhead Wilson*, in which the protagonist, a likeable oddball, becomes fascinated with fingerprints. Through their use in a trial proceeding he solves a murder mystery and brings to light a baby switching that had occurred many years in the past. Fingerprinting was not a new technique at that time. It had been developed earlier in the century as a method of identity verification in India by a British official, William Herschel. Its utility as a tool for identifying criminals soon became evident, and it gained use in law enforcement simply as a means of identifying those who had been arrested or incarcerated. It was not until 1910, however, that fingerprint evidence was introduced into trial proceedings. Mark Twain was in this respect, as in so many others, ahead of his time.

The identification of individuals by comparison of their fingerprints with stored records was deemed very reliable, in part because the file and test fingerprints could be of good quality, typically taken of all ten fingers (the so-called ten print) by a person trained in the process. The essential idea behind the method was that fingerprints are unique to each individual; no two people, not even identical twins, have exactly the same fingerprints. Furthermore, it was held that a person trained in fingerprint comparisons could unambiguously associate a fingerprint with a stored print.

In the first American criminal trial involving fingerprint testimony, Thomas Jennings was convicted of the murder of Clarence Hiller, largely on the basis of fingerprint evidence. The Hillers had recently painted their house, and four fingers of a left hand had been imprinted in the still-wet paint on the back porch. Four experts for the prosecution testified that in the course of their work for various law enforcement agencies, they had looked at thousands of sets of fingerprints, and that the prints in the paint were undoubtedly those of Thomas Jennings. On appeal of his conviction Jennings argued that the fingerprint evidence was inadmissible, but his conviction was upheld. The court found that there was a sound scientific basis for fingerprint identification, and that the courts could take judicial notice that the method in was general and common use.

The court took for granted that whatever reliability attached to the use of fingerprints in identification contexts would carry over unproblematically to the use of fingerprints as evidence in criminal trials. The question of just what it took to make a match was never asked, let alone answered:

> What was striking in *Jennings*, as well as the cases that followed it, is that courts largely failed to ask the difficult questions of the new identification technique. Just how confident could fingerprint identification experts be that no two fingerprints were really alike? The witnesses in *Jennings* claimed to have personally looked at several thousand fingerprints—but even if the several thousand were all different, that hardly proved that *all* fingerprints are different.
>
> Moreover, when evaluating prints in the context of criminal identification, examiners had access to full sets of ten complete prints; in the forensic context, by contrast, they typically had fewer, often only one. . . . Neither the court in *Jennings* nor subsequent judges ever required that fingerprinting identification be placed on a secure statistical foundation. How many similarities, how many points of resemblance, between two prints were necessary in order to declare a match? . . . To this day, fingerprint examiners have never been required to give an answer to this question, and to this day, they remain incapable of providing an answer that is rooted in a persuasive statistical model of fingerprint variation. (Mnookin 2001b)

Notwithstanding these faults, fingerprint identification quickly became a widely used forensic tool, vested with essentially unassailable authority. Jennifer Mnookin reports that she and fellow researcher Simon Cole have been unable to document a single case in the years following *Jennings* in which fingerprint evidence was declared inadmissible by a trial judge (Mnookin 2001b, 23n30). The fingerprint experts that testified in the early years often established their expertise by highly persuasive in-court demonstrations (Cole 2001, 190–94). Typically, the expert had each of the twelve jurors make imprints on two pieces of cardboard while the expert was out of the room. The expert then developed the prints and successfully paired them all. Appellate courts found that these demonstrations merely confirmed what was clearly evident, "to illustrate the methods of the system of finger-print identification and the truth of the claim that invisible finger prints can be developed and identity of the maker revealed by a simple process to positive certainty" (*Moon* 1921).

These in-court demonstrations went a long way toward solidifying the reputation of fingerprint identification as a forensic tool nonpareil. Fingerprint experts testified to what they considered matters of certainty, not opinion, and no one successfully challenged them. Before long a professional organization for fingerprint examiners, the International Association for Identification (IAI), had been formed. As the use of fingerprint identification grew in criminal cases, the aura of infallibility of the method was carefully nurtured. By 1980 the IAI had made it a matter of professional misconduct for a fingerprint examiner to provide courtroom testimony that labeled a match "possible, probable or likely" rather than "certain," other than under threat of a contempt citation (International Association for Identification 1980).

Thus, fingerprint identification came to have a virtually unassailable quality in the courts. In some senses this seems anomalous, because during this same period many forms of expert testimony in general were regarded with considerable skepticism:

> In this context, fingerprinting seemed to offer something astonishing. Fingerprinting—unlike the evidence of physicians, chemists, handwriting experts, surveyors or engineers—seemed to offer the kind of solid, indisputable evidence that was hoped for from science. Writers on fingerprinting routinely emphasized that fingerprint identification could not be erroneous, or that its results were both consistent and certain.
>
> Unlike so much other expert evidence, which could be and generally was disputed by other qualified experts, fingerprint examiners seemed always to agree. Generally, the defendants in fingerprinting cases did not offer fingerprint experts of their own. Because no one challenged in court either fingerprinting's theoretical foundations or, for the most part, its actual operation in any particular instance, the

technique came to seem especially powerful. Fingerprinting therefore offered precisely the kind of scientific certainty that judges and commentators, weary of the perpetual battles of the experts, yearned for. (Mnookin 2001b, 38–39)

Fingerprint identification became the exception in part because it had become a guild activity. The field had little or no application outside law enforcement, so there was not a wider marketplace in which to practice. Fingerprint identifications were virtually never contested by expert testimony that might befuddle the jury or call into question the reliability of the identification. The field developed practices that served to solidify its place in the courtroom, such as a requirement that any fingerprint identification offered in testimony had to have been verified by a second or even a third examiner. Notably, however, these verifications did not consist of blind tests on the same print, but rather were corroborations carried out by a person who usually already knew the conclusions reached by the first expert. The courtroom demonstrations already mentioned also served to build confidence in the veracity of the testimony, even though they in fact demonstrated nothing about the potential error rates in identifications of the latent prints, often of poor quality, that turned up in real cases. The potential for abuse of the special status of fingerprint expertise was illustrated in 1992 when investigators uncovered widespread fingerprint fabrication by New York State police officers in some forty cases over a period of eight years. None of the fabrications was challenged by the defense, though it turned out that they were poorly done and could have easily been detected (Cole 2001, 274–81). More recently, the Los Angeles times reported on the contents of an internal report of the Los Angeles Police Department, in which it was acknowledged that people have been falsely implicated in crimes because the department's fingerprint experts wrongly identified them as suspects (Rubin and Winton 2008).

If there were not reasons enough to challenge fingerprint identification before *Daubert*, there certainly were in the wake of that decision. Nevertheless, the first challenge to fingerprinting that cited *Daubert* did not occur until 1999, fully six years after the *Daubert* decision (*United States v. Mitchell* 1999). Fingerprint identification portrays itself as a scientific practice, but in fact it lacks certain critical characteristics of science. In the first place, one of its core hypotheses, that every person has a unique fingerprint, has not been tested. Indeed, this is not a provable hypothesis. It might be possible to show by some convincing sort of test that every one of the more than forty million sets of prints held in the FBI files is different from every other, but that has not been done, and no one really knows how it could be with current technology. The best that can be said is that individual prints differ in various ways, and these differences can be exploited to make identifications, in the best cases with high reliability.

Secondly, a key consideration in the *Daubert* opinion is that account should be taken of the potential error rates in the methods that form the basis of expert testimony. The error rate in fingerprint identification is a hotly contested issue. The fingerprint identification community has held that the method, properly applied, is error-free. This claim is patently absurd, particularly in light of the difficulties in making identifications of latent prints, sometimes of only one finger, sometimes smudged or lifted from an irregular surface. The practices within the fingerprint identification community virtually foreclose any real testing of error rates in practice. Experiments with identification comparisons using blind or double-blind testing have never been carried out. The evidence adduced from various sources, however, suggests that misattributions of fingerprints are likely quite common (Haber and Haber 2003, 339–59). For example, the IAI, through its Latent Print Certification board, offers certification and recertification to latent print examiners. The examiners who take the certification test must have had both formal training and experience on the job. To pass the fingerprint comparison part of the test, the test taker must identify correctly twelve or more of fifteen latent prints by comparison with the ten-print cards, without making a single false identification. Failure on this part of the test accounts for nearly all certification failures. Between 1993 through 2001, the pass rate on the certification test hovered in the vicinity of 50 percent. Certification is not a requirement of employment as a fingerprint examiner, and a majority of those in the profession never take this test. Given that those who do apply for certification are likely to be among the more active and professionally involved fingerprint examiners, the claim that fingerprint identifications are made reliably by individual examiners is simply not credible.

Elizabeth Loftus and Simon Cole have recently recounted a number of specific examples of fingerprint misattribution. A highly publicized recent example is that of Brandon Mayfield, an Oregon attorney and Muslim convert who was held for two weeks as a material witness in the Madrid bombings of March 11, 2004, a terrorist attack in which 191 people were killed. A bag found in Madrid after the explosions contained detonators and explosives. The Spanish authorities could not identify a latent fingerprint on the bag, and they e-mailed it to other police authorities. A senior fingerprint examiner in the FBI identified Mayfield as the source of the latent print. Mayfield was in the database because of his military service and a 1984 arrest for burglary. The identification was validated by a second independent fingerprint expert. Two weeks later the identification was retracted, however, and the FBI issued an apology to Mayfield. (The Spanish police had attributed the print to an Algerian national living in Spain.) In another example, Stephan Cowans was released from prison in Massachusetts after serving six years for the nonfatal shooting of a police officer, after DNA evidence demonstrated that he could not have been the guilty person. He had been convicted on the basis of both eyewitness and crucial fingerprint evidence (Loftus and Cole 2004, 959).

Many post-*Daubert* challenges to fingerprint identification testimony have been mounted, but until relatively recently none have been successful. As an example, in *United States v. Havvard* (2000), the defendant sought to bar the government from offering opinion on whether a latent fingerprint recovered from one of the firearms in question matched Havvard's left index figure, contending that such opinion evidence does not meet the standards of reliability for admissible expert testimony under *Daubert* and *Kumho Tire*. The court published a rather detailed analysis of the issues and concluded that "the court believes that latent print identification is the very archetype of reliable expert testimony under those [*Daubert* and *Kumho Tire*] standards."

In 2002, federal district court judge Louis Pollak caused a great furor when he ruled quite differently in *United States v. Plaza* that fingerprint experts could not testify that a suspect's prints definitely match those found at a crime scene. Judge Pollak took judicial notice of the premise that every individual has unique fingerprints of lifetime permanence, and he allowed testimony on the methods employed in identifying the similarities and dissimilarities between the prints so compared. He also allowed that defense experts would be permitted to rebut prosecution evidence. He drew the line by ruling that the experts were not to be permitted to testify that the questioned latent print and the known inked print were produced by the same digit. An opinion to this effect, the judge thought, was too subjective in nature, given what he saw as limited evidence of the method's reliability. In his view, practitioner error had never been adequately quantified. On further consideration, however, and after hearing presentations from both prosecution and defense experts, Judge Pollak reversed himself and allowed the introduction of fingerprint identification testimony. He did make it clear, though, that he viewed fingerprint identification as a form of technical expertise, and not science.

My purpose in relating all this information about fingerprint identification is not so much to judge arguments for or against its reliability, nor to critically evaluate its claims, but rather to point to it as an exemplar. Although it has to some extent lost its reputation for near infallibility, for nearly a century fingerprint identification has enjoyed an unprecedented level of authority as courtroom testimony. What accounts for the ascendance of this method to such a position, and what has sustained it, even in the face of more aggressive challenges recently?

*Fingerprint identification carried an aura of scientific rigor.* It came into common use in an era when there was great interest in measuring human anatomical features. Among those who first studied fingerprints, the eugenicist Francis Galton is notable for developing a classification system. His work and that of others lent an air of scientific interest to fingerprints as objects of study. It also helped establish the credibility of the idea that no two persons had the same fingerprints. Mark Twain's *Pudd'nhead Wilson* illustrates how the claims of fingerprint identification for reliability and discriminatory power could come to be accepted without serious questioning.

*Fingerprint identification offered visible objects for inspection and judgment.* Visible displays of latent fingerprints lifted from crime scenes, coupled with the test prints against which comparisons are made, offer an element of immediacy and concreteness that instills confidence in the claims made for the identifications. In the minds of the judge and jury, fingerprint identification is about visible, real things, as opposed to opinions about past states of peoples' minds, probabilities based on statistical analysis of databases, and other forms of expert testimony that deal with more abstract concepts.

*Fingerprinting made a strong claim to incontrovertible knowledge.* As fingerprinting gained acceptance, and fingerprint identification became an accepted profession, an ethic of sorts developed: experts would not disagree with one another in the courtroom on identifications. Instead, in normal practice, identifications offered in testimony has already been "verified" by another examiner. It is not surprising, then, that identifications made according to a prescribed set of procedures, by experts who had been thoroughly trained in the methods of identification, were accepted as certain. Judges were strongly attracted to this claim; any evidence or testimony that could lead to conclusions formed with certainty, or something very near it, were much to be valued over notoriously unreliable eyewitness testimony, or the conflicting testimonies of experts on opposite sides of an issue.

*Fingerprint identification experts formed a consensual social community.* This was a vital aspect of the presentations made by fingerprint examiners. Formation of professional societies, particularly the IAI, ensured that the practitioners operated under a common set of procedural rules and standards of work. The community of examiners, in company with the law enforcement organizations within which they were embedded, recognized unanimity as the way to maintain a high level of authority for fingerprint identification, and the way to maintain unanimity was to allow only conservative interpretations of a match. It is not entirely clear why there did not arise a corps of examiners operating outside the law enforcement community who could serve as countervailing expert testimony for the defense. In part it was simply that the fingerprint evidence was generally quite convincing. Further, it was not evident how to go about challenging it, or how to bring together the needed resources for doing so. In any case, the testimony of the prosecution's fingerprint expert was virtually never challenged.

Because fingerprint identification is about the study of objects that exist in the natural world, and is systematic in its characterization of them, it might seem reasonable to think of it as a form of scientific study. We have seen, however, that in practice it falls short of being scientific. There have been no scientifically motivated studies of what aspects of comparison best make for identification, no credible outside-party studies of the error rates made by examiners on real samples, nor any tests of comparison between different examiners when double-blind studies are carried out. Fingerprint identification requires no scientific training. Rather, training consists in learning how to make detailed comparisons of prints, a cognitive skill that may require accumulation of a great deal of tacit knowledge and experience, but not one that rests on a broad understanding of underlying aspects of the physical world. For all these reasons, the world of science abhors the idea that fingerprint identification (and related forensic methods such as hair sample analysis and ballistics measurements) should be thought of as science, and has not been shy about saying so (Kennedy 2003).

By contrast, the much newer technique of DNA identification, which often competes with fingerprint identification in the courtroom, is dependent on current scientific theories in biology. Its claims for authority rest on the plausibility of the story it tells of the origins and significance of DNA, and the complex procedures and analyses involved in DNA analysis. Its initial reception in the courtroom in some ways paralleled that of fingerprint identification, but it has come into its own in a much different environment with respect to admissibility of expert evidence.

<br>

## DNA IDENTIFICATION

I began this book with the story of Eddie Joe Lloyd, exonerated and released from prison after having been proved innocent of a murder for which he served over seventeen years. His exoneration was based on the results of DNA testing. As I write this, the Innocence Project reports that 123 persons have been exonerated after conviction of serious crimes such as rape and murder on the basis of DNA testing of crime scene evidence, and the number continues to grow. DNA evidence also serves in many other capacities in the legal system. For example, it has been employed to settle claims of patient sample or nursery mix-ups in hospitals, in paternity cases, and in immigration disputes. DNA-based genetic profiling was used to identify the remains of the many victims of the September 11, 2001, attacks on the United States. A famous instance of its use involved a Sri Lankan infant, identified only as "Baby 81," separated from his parents when a huge tsunami swept over Sri Lanka in December 2004 (Associated Press 2005). After being rescued from under a pile of rubbish, the baby was claimed by several couples, but the true parents were confirmed by DNA testing.

The use of DNA for identification and other purposes within the legal system is quite obviously the result of basic scientific research. James Watson and Francis Crick's famous model of the structure of DNA (or deoxyribonucleic acid) gave rise to a cornucopia of molecular biological findings that continue to shower the world with new insights and applications. DNA, a gigantic molecule found in the nuclei and mitochondria of cells, is the carrier of the genetic information that gives rise to the whole organism. The sequences of the chemical units, known as bases, in the long DNA string provide the instructions for building and running each and every cell in the body. The three billion base pairs that form the entire string of DNA within the nucleus constitute the human genome. Any two humans have in common the vastly greater part of their DNA; only about one-tenth of 1 percent of the genome differs from person to person, and that variation is found in only about 2 percent of the entire genome. Nearly all that variation is heritable; the DNA that children inherit from their parents is a combination of a strand from each parent, so children are genetically linked to their parents. Every cell in the body, whether from saliva, skin tissue, blood, hair, or bone, contains the same DNA, which remains essentially unchanged throughout life.

The idea of using DNA to identify individuals, or for other legal purposes, grew naturally out of earlier work involving heritable variations in blood types. The various human blood cell types (O–, A+, etc.) are determined by variations in substances that occur on the surfaces of red blood cells. Knowledge of these blood types has been used to serotype those needing blood transfusions. Similar matching methods are used in cross-matching donors and recipients of bone marrow or whole organs for transplantation. These earlier matching techniques found their way into the courtroom in connection with paternity and other sorts of cases. DNA, with its vastly greater level of specificity for individuals, was an obvious place to look for more precise means of identification. No two persons, other than "identical twins" (or identical triplets, etc.), have exactly the same DNA.

Using chemical tools that act like molecular scissors, the long DNA strand can be clipped into shorter sequences. Many complicated considerations that need not concern us are involved in analyzing the shorter lengths. Suffice it to say that variations in the shorter lengths (called alleles), and in the sequences of bases within those alleles, provide ample basis for distinguishing one sample of DNA from another, and for identifying the sample with a specific person when a sample from that individual is available for comparison (Coleman and Swenson 1994, chap. 3).

The first use of DNA in a criminal case occurred in England in 1987. Alec Jeffries, a professor at the University of Leicester, studied the DNA patterns from two separate rape/homicide crime scenes, and compared them to those of a confessor of one of the crimes. His findings excluded the confessor as the source of the DNA, and strongly suggested that the same individual committed the two crimes. The DNA samples from the crime scenes were compared to those collected from adult males living in the region, who were encouraged to come forward for voluntary blood donation (Wambaugh 1999). In time the single perpetrator of both murders, a baker named Colin Pitchfork, was identified. Shortly afterward, in 1989, Gary Dotson became the first person in the United States to have a conviction overturned on the basis of DNA evidence, which excluded him as the perpetrator. He had served eight years of a twenty-five-to-fifty-year sentence for rape.

Often DNA testing must be carried out on very small samples: a tiny drop of blood, spittle, skin cells scraped off the perpetrator by the victim, a few hair roots. To obtain sufficient material to carry out effective testing, the extremely tiny amounts of the DNA strands of interest must be "amplified" to make more sample. This is done using the polymerase chain reaction (PCR), invented in 1986 in Nobel Prize–winning work by Kary Mullis. Use of PCR has made it possible to adduce evidence derived from extremely unpromising samples. This powerful method does, however, come at a price: extreme care must be taken to prevent contamination by foreign biological materials in the course of collecting and preserving a sample. What this means in general is that the techniques and procedures employed in forensic processing of DNA materials are very demanding.

DNA identification techniques got off to a promising start with Jeffries's work. He dubbed the method he had developed "DNA fingerprinting," a gesture toward the long-established and widely accepted methods of fingerprint identification. DNA methods were rapidly taken up by forensic laboratories in the United States and elsewhere. For example, George Kaufman, when confronted with DNA evidence in 1989, confessed to raping sixteen women over a four-year period in Melbourne, Australia. This auspicious beginning might thus be taken as another instance of the successful exercise of scientific authority. But it quickly became evident that the new technology was not going to have an easy time of it in court, nor, for that matter, in the scientific literature (Lynch et al. 2008).

On February 5, 1987, Vilma Ponce and her two-year-old daughter were stabbed to death in the Bronx. Jose Castro, a handyman in the neighborhood, was questioned. The police noticed a bloodstain on his watch. Samples taken from the deceased victims and the watch were sent to a commercial laboratory that performed DNA analyses, which reported a match between crime scene samples and the bloodstain on the watch. No mention was made in the report of any difficulties or ambiguities in the processing or analysis. The prosecution moved to have the results of this testing admitted as evidence, and the defense challenged.

There followed a twelve-week period of pretrial hearings in the Superior Court of Bronx County, New York, on the admissibility of the DNA evidence. In the end, following the testimony of expert witnesses on both sides, judge Shira A. Scheindlin found that DNA identification tests were acceptable in general, but that in the specific case at hand the laboratory had not followed approved procedures; the DNA evidence of guilt was not admitted (*People v. Castro* 1989). Recall that this case occurred pre-*Daubert*; most courts were still applying the *Frye* standard of general acceptance in the relevant community. After an unusually thorough study of the underlying scientific issues, Judge Scheindlin accepted the soundness of the science underlying DNA testing, as well as the particular procedures employed in analysis of the samples. In this sense, the authority of science was upheld by the court. The judge, however, found that Lifecodes, the commercial laboratory that had processed the samples, had not performed certain experiments nor carried out certain controls necessary to produce a reliable result (Patton 1990; Coleman 1994, chap. 1).[2] Shortly afterward, in Minnesota, in *State v. Schwartz*, the court again ruled against admissibility of DNA evidence on the grounds that the tests had not been properly conducted. These cases highlighted the challenges that lay ahead to the use of DNA evidence in criminal proceedings (Harmon 1990, 149).[3]

DNA technologies have advanced considerably in power, sensitivity, and reproducibility. To this day, however, there remain controversies relating to estimating the probabilities associated with matching of DNA samples (as will be discussed below). Perhaps more important, the possibilities of laboratory error and bias have become the weakest links in the chain of argument for DNA evidence (Koehler 1997; Lazer and Meyer 2004, 377–78; Bieber 2004, 38–41; Thompson 1997). In response to the concerns that arose in cases such as *Castro* and *Schwartz*, Rule 702 of the Federal Rules of Evidence was modified in 2000 to include a requirement for a foundational showing that "the witness has applied the principles and methods reliably to the facts of the case." Broadly speaking, since the early 1990s the emphasis has shifted from questioning the basic science to challenging the fairness, impartiality, competence, and integrity of the criminal justice system itself. William Thompson, a law professor and former defense attorney with experience in cases involving DNA, has this mordacious view of matters: "Thus, forensic DNA testing is not an independent area of academic science; it is a technical field in which procedures are designed, shaped and performed specifically to play a role in litigation. Moreover, the field is dominated by experts and laboratories whose primary clients are law enforcement agencies, and whose typical role in litigation is to support evidence in support of criminal prosecution. In this context, pressure exists to distort science to serve law enforcement goals" (1997, 408). The O. J. Simpson trial in 1995 furnishes a highly publicized example of the kind of courtroom atmosphere that can prevail. The prosecution's DNA evidence was

2. Castro later confessed to the murders.

3. As an aside, it should be noted that a positive identification of a match of two DNA samples is generally much more difficult than an exclusionary or negative match. That is, it is much easier to conclude that two DNA samples are *not* from the same person than it is to conclude that that they are. Thus, the use of DNA comparisons to secure post-conviction exonerations does not generally meet with the same challenges that attend attempts to introduce DNA identifications in trials.

undermined by defense allegations that the Los Angeles Police Department handled the samples incorrectly, that there was a significant possibility of lab error, and that there might have been a conspiracy to frame Simpson (Mueller 1996; Thompson 1996; Linder 2000; Nowak 1994). The jury was led to distrust the law enforcement system, not DNA identification per se.

I have already indicated that statistical and probabilistic considerations are very important in drawing conclusions from the experimental data in DNA identification procedures, as well as for their effects on judges and juries. Any positive DNA match must be stated in terms of a probability that the crime scene sample and test sample are from the same person. For example, it might be estimated that there is a one-in-ten-million chance that a particular set of DNA patterns derived from material taken from the defendant and the corresponding patterns taken from the crime scene belong to different persons. The factors involved in making such an estimate are quite involved. They include, among other things, the quality of the DNA patterns elicited in the analysis procedures (which depends on quality of the initial samples), the number of DNA strands (so-called loci) being compared, and the particular experimental methods employed in sample processing. In addition, the ethnic or racial backgrounds of suspect and victim figure into estimating the probabilities associated with a match. In the early years of DNA typing, a controversy played out in the pages of the journal *Science* regarding the appropriate methods for calculating the probabilities of a match, because of prospects that ethnic subgroups might maintain genetic differences through endogamy (Lewontin and Hartl 1991; Chakraborty and Kidd 1991; C. Wills 1992). Members of relatively self-contained subpopulations share DNA with more points of similarity than with DNA from other population groups. An NRC panel on DNA forensics was appointed to study the various issues related to DNA typing, including the role of population genetics. The report, issued in 1992, itself produced controversy because of certain simplifying recommendations relating to calculations of probabilities of matching (NRC 1992, 82–91; Devlin, Risch, and Roeder 1993). In short order a second NRC committee was formed to study this particular issue (NRC 1996, 65–73). The new report, issued in 1996, went far toward answering many of the objections made to its predecessor, but some in the scientific and legal communities found it lacking in significant respects (Morton 1997; Thompson 1997).

Finally, there is the question of how well the judge and jury understand the probabilistic and statistical considerations that go into a proper assessment of the likelihood of a DNA match. Assuming no laboratory error limitations, what is the likelihood that DNA from the crime scene and a DNA sample identified with a suspect will form a match if the two samples arise from two different individuals? Because most people are not sophisticated about such matters, the results of DNA analyses can be presented in various ways to influence jurors' understanding of their significance (Koehler 2001; Kaye and Koehler 2001; Lindsey, Hertwig, and Gigerenzer 2003). We will return to the issue of statistics and its role in assessment of scientific evidence in the courtroom in the next section, dealing with tort litigation.

The story of DNA profiling is instructive in showing the power of science to exercise authority, although it also illustrates the many factors that can operate to limit or even negate that authority. It is useful to look at the history of DNA profiling in light of the four factors identified above that made for the authority of fingerprint identification.

*It has an aura of scientific rigor.* Unlike fingerprint identification, DNA profiling is an outcome of basic scientific discovery. It therefore has from the beginning carried a scientific imprimatur. Whereas the claims of fingerprint identification to being somehow scientific were eventually rejected, DNA fingerprinting technology is firmly grounded in molecular biological sciences such as genetics and cell biology. It was just this unambiguous descent from basic science that lent DNA profiling its great authority at the outset, and has sustained it through periods in which it has been challenged in the legal arena.

*It deals with visible objects, readily seen and comprehended.* Here the comparison with fingerprint identification is close. The material shown as evidence in the two cases are formally quite comparable: on the one hand, fingerprint images to be compared, and on the other, developed films, called autorads, that show, for example, comparisons between band patterns for suspect and crime scene sample. The DNA profiling technique is complex, however, and the story of how the forensic evidence comes to be in the form shown, as well as the significance of the patterns, is less readily explained than for fingerprint evidence.

*The expert testimony is largely consensual.* The culture of consensus among fingerprint examiners with respect to evidence presented in the courtroom contributed enormously to the credibility of fingerprint identification. Defense counsels often had no realistic means of challenging the testimony of the experts. A comparable environment of consensus in the community of experts has not existed for DNA-related testimony. Within just a few years of the introduction of DNA profiling evidence into criminal cases, successful challenges to its admissibility were mounted, as in the *Castro* and *Schwartz* cases. A key expert testifying for the defense in the *Castro* case was Eric Lander, a well-known molecular geneticist, founding director of the Broad Institute, and a professor of biology at MIT.

In 1990, the case of *United States v. Yee* was widely expected to be a rigorous rite of passage for DNA testing (Coleman 1994, chap. 1). Yee and two others were charged with murdering David Hartlaub; DNA evidence adduced from blood samples was at the heart of the prosecution's case and was vigorously challenged by the defense. The prosecution called six expert witnesses, including Thomas Caskey of the Baylor College of Medicine, who had served as head of a panel studying DNA typing for the Congressional Office of Technology Assessment, and Kenneth Kidd, a Yale geneticist. The defense experts included Richard Lewontin, a well-known Harvard geneticist, and Daniel Hartl, a geneticist at Washington University. At about this time Lewontin and Hartl (1991) wrote a controversial paper in *Science* that challenged the then-prevalent methods of calculating the probabilities associated with DNA matching, on the grounds that ethnic subgroups within major racial categories exhibited genetic differences that are maintained because of extensive intermarriage within those ethnic populations. The defense lost this particular challenge to the admissibility of DNA evidence, but testimony offered in the process spawned doubt about the authenticity of DNA methods, even when it gained admission into testimony. The controversy generated by this case and the Lewontin-Hartl paper led eventually to the appointment of the first NRC panel.

It is a given within science that there will not be unanimity of outlook on many topics among those exploring the leading edge of a scientific field as it develops. Indeed, as Polanyi has remarked, the tension created by working within accepted frameworks of knowledge, while seeking to break out of those frameworks, is a core characteristic of scientific research practice (discussed in chapter 4). But this fact leads to a dilemma: in order for any subgroup within the scientific enterprise, or for science more generally, to exercise credible authority outside the scientific community on a given topic, there must be an appearance of consensus on the key elements.

Recognition of scientific authority by society is deemed important within the scientific community. To illustrate, Daniel E. Koshland Jr., the editor of *Science*, writing in 1994 (a time, as we have seen, when DNA typing was being vigorously challenged), called attention to recent successes of DNA typing, and went on to say:

[The growing] acceptance of the validity of DNA evidence is exactly what most scientists in this area have believed appropriate, and a rebuke to the judicial process that has been so slow to accept DNA evidence by failing to see that a couple of outspoken individuals were less representative of the scientific community than the vast majority of careful scholars. (It is notable that the scientists prominent in casting doubt on DNA use for the prosecution seem to be nowhere in evidence to cast doubt on its use for the defense.) One note of caution that the doubters raise was correct: the need for careful analyses on well-authenticated samples. But their argument that elaborate state machinery is needed to monitor the work of testing agencies is clearly overkill. Undergraduates are now doing good DNA tests, and their results can easily be checked by standard control samples. (1015)

There are two striking aspects of these comments by a distinguished biologist and editor of one of the world's most widely read scientific journals. First, Koshland implies that jurists should be able to make judgments about scientific disputes as they apply to legal issues, even when well-known and esteemed scientists are raising issues that would certainly seem abstruse to a nonexpert. Secondly, Koshland exhibits a remarkable nonchalance about what needs to be done to establish and safeguard the integrity of the processes of evidence formation. On this note we come to the fourth and final aspect of establishing authority.

*DNA testing makes claims to virtually incontrovertible knowledge.* The basic science underlying DNA typing leads to the following conclusion: The probabilities that two DNA patterns formed from a sufficient number of loci will match if the DNA is from two different persons (other than identical twins) can be calculated conservatively to be extremely small—so small, in fact, as to constitute a virtual certainty that only one person can have given rise to the matching DNA. We have seen that this broad claim has been attacked, or its power to persuade eroded, by several factors, some of which have to do with shortcomings in the procedures employed in the analysis. Other attacks have concerned certain aspects of the underlying science, particularly the calculation of the probabilities associated with a match.

Life scientists with interests in this area have actively combated what they perceive as attacks on the authority of science. For example, in 1994, at the time of a great public interest in DNA typing occasioned by the

O. J. Simpson trial, two scientists who had been prominent in arguing on opposite sides of various controversial aspects of DNA typing wrote a paper in the journal *Nature*, in which they attempted to lay to rest many of the doubts that had arisen. Eric Lander (who, as mentioned above, had testified for the defense in the *Castro* case) and Bruce Budowle, head of the FBI's Forensic Science Research and Training Center, admitted at the outset that there had been controversy regarding DNA typing, and that they had been on opposite sides on certain issues. They argued, however, that the "DNA wars" were now over. Addressing the concerns of some scientists relating to population genetics, they argued that, in spite of academic disagreements about the assumptions made in estimating probabilities of matches, the methodology being implemented was extremely conservative. They commented on the relationship of science and the legal system as follows:

> It is easy to forget that this new debate is purely academic. The most extreme positions range over a mere two orders of magnitude: whether the population frequency of a typical four-locus genotype should be stated, for example, as $10^{-5}$ or $10^{-7}$. The distinction is irrelevant for courtroom use.
>
> Rehashing issues may be a harmless pastime in the academic world, but not so in the legal system that lives by the dictum *stare decisis* (let the decision stand). From the standpoint of law enforcement, it is better to have a settled, if slightly imperfect, rule than ceaselessly to quest after perfection. . . .
>
> Most of all, the public needs to understand that the DNA fingerprinting controversy has been resolved. There is no scientific reason to doubt the accuracy of DNA typing results, provided that the testing laboratory and the specific tests are on a par with currently practiced standards in the field. The scientific debates served a salutary purpose: standards were professionalized and research stimulated. But now it is time to move on. (Lander and Budowle 1994, 738)

Budowle and Lander's attempt to put to rest lingering doubts about the efficacy of DNA typing was not wholly successful. Even as they were writing, the NAS was convening a second panel to prepare an update on the 1992 DNA *Technology in Forensic Science* report. Most particularly, the panel was charged with a reexamination of the issues tied to population genetics. As we have already seen, the second committee's report was received much more favorably than the first committee's. But aside from the details of how the probabilities are best calculated, the basic scientific assumptions and the analytical methodology of DNA typing were essentially taken for granted. It was the correct *application* of those basic scientific principles in the forensic laboratory that was then, and has continued to be, subject to questioning in the courtroom.

Budowle and Lander's aim was to play down the population genetics issue. They correctly sensed that the controversy over the methodological issue of how to properly account for population genetics in the analysis of DNA results was having an adverse affect on the authority of DNA typing evidence. If there is disagreement among scientists who should be most deeply knowledgeable about a matter that seems central to the method, the authority with which the results of DNA typing can be asserted is diminished.

In a similar but less direct way, doubts about the quality of DNA typing practice detract from the aura of authority associated with the method. In this regard, DNA typing has had a much different history in its early stages than fingerprint identification. Springing as it has from the basic life sciences laboratory into the courtroom, DNA typing has carried with it some of the messiness of unfinished scientific business. As DNA methodologies have steadily improved, however, the underlying science has been more confidently asserted in advancing the efficacy of the method. In addition, strict protocols for sample handling and analysis, as well as certification of technical personnel, have strengthened the credibility of the proffered testimony (Saks 2005a; Lynch 2003). The history of DNA typing shows us how the authority of science in the courtroom is dependent on both a consensus in the scientific community and, concomitantly, a solid body of authoritative basic science (Lynch et al. 2008). It also reveals a fundamental limitation in the extent to which scientific authority can operate in the law: "DNA test results are embedded in the details of criminal cases that will always, and should always, involve public considerations of the singular elements of a case. The meaning and significance of 'scientific' testing cannot be divorced from such consideration" (Lynch 2003, 97).

## TORTS

In closing this chapter I want to return to the subject of tort law, because it is here that the authority of science has been most severely challenged, and where the battle lines, for that is what they are, remain scenes of active combat. The animating purpose of tort law is to fix blame, to determine whether and in what degree there has been a breach of the duty of care. Part of the difficulty with evaluating the role and influence of science in tort law is that tort cases are laden with complex issues of social policy, attitudes of everyday folk toward government and large corporations, and a common tendency to fix blame for tragedies that are in fact no one's fault. In the course of the twentieth century, the idea came into currency that there should be some form of justice for individuals caught up in the workings of a modern capitalist society and suffering mishaps for which could not be blamed.

Product liability suits emerged as a branch of tort law, and with them the doctrine of strict liability, so that over time legal philosophy shifted to see tort law as a policy instrument. As Sheila Jasanoff has written, "The view that the primary function of tort litigation was to compensate injured persons gained ground over the countervailing doctrine that torts were mainly a legal device for punishing fault" (1995, 27). But this change in outlook, to the extent that it found solid footing, created an entirely new kind of arena in which science was invoked in the courtroom: "The liberalization of products liability law to compensate more victims encouraged litigation based on uncertain causal theories and drew seemingly insoluble scientific controversies into the courtroom. Put differently, a change in judicial policies for resolving social uncertainty—who should bear the cost of accidents—provided the impetus for addressing new kinds of scientific and technical uncertainty in court" (31).

## AGENT ORANGE AND THE CONCEPT OF CAUSATION

Massive increases in the size of tort suits, involving in some cases hundreds of thousands of plaintiffs (as in silicone breast implant litigation), occasioned entirely predictable reactions from defendants (Hooper, Cecil, and Willgang 2001). This is not the place to attempt a summary of the vast literature dealing with the various aspects of tort litigation, nor to attempt a retelling of the stories behind the most famous cases.[4] The *Daubert* case functioned in the context of this highly contentious atmosphere as an attempt to set down key considerations that should operate in determining

---

4. Among the books that deal with various aspects of the controversies are Angell 1996; Jasanoff 1995; Huber 1993; Foster, Bernstein, and Huber 1993; Foster and Huber 1999; and Green 1996.

whether and when scientific testimony is admissible. Prior to the bevy of suits over the effects of Bendectin, which included the *Daubert* suit, in the 1980s there had been a huge number of suits related to the effects of Agent Orange, a defoliant used extensively in the Vietnam War. Agent Orange had allegedly caused a variety of illnesses, including cancer and neurological disorders, in Vietnam War veterans, and had been allegedly responsible for birth defects in their offspring. The primary offending component of Agent Orange was dioxin, a contaminant formed in manufacturing the defoliant itself. Establishing causation in such cases was virtually impossible, given the wide range in the timing, levels, and conditions of exposures of the plaintiffs, as well as the passage of considerable time. Judge Jack Weinstein oversaw a $180 million settlement on behalf of the six hundred thousand veterans included in the Agent Orange class action, and then turned to the claims of several hundred plaintiffs who had opted out of the class action (Green 1992).

The defendants, a group of chemical manufacturers, moved for summary judgment on the grounds that plaintiffs had failed to establish a causal connection between their exposure to Agent Orange and the various diseases which they attributed to that exposure. In a lengthy opinion, Judge Weinstein granted the defendants' motion. His several views on what constituted acceptable scientific evidence have been well summarized by Michael Green (1992) and Peter Schuck (1986, 234–41). Weinstein's opinion was later influential with judges trying the Bendectin cases (*Agent Orange Product Liability Litigation* 1985). In brief, he strongly endorsed the efficacy of epidemiological studies: "A number of sound epidemiological studies have been conducted on the health effects of exposure to Agent Orange. These are the only useful studies having any bearing on causation." He was dismissive of animal studies: "[They] are not helpful in the instant litigation because they involve different biological species. They are of so little probative force and are so potentially misleading as to be inadmissible." Finally, he also expressed deep skepticism regarding much of the testimony of the plaintiffs' expert witnesses, whose opinions he perceived to be outside the mainstream in their areas of specialty.

Because Judge Weinstein enjoyed wide recognition as a trial judge, teacher, and legal scholar in matters relating to evidence, his handling of the Agent Orange case set a powerful precedent. His reasoning and judgment in assessing the plaintiffs' evidence has been questioned, however, in part because of his uncritical acceptance of three epidemiological studies that might have been scrutinized more thoroughly for relevance and the presence of methodological errors (they were not, however, challenged by the plaintiffs' attorneys), and for his quick dismissal of other forms of evidence (Green 1992; Cranor 2006, 224–27, 248–52). The same disposition to favor epidemiological studies over other forms of evidence relating to causation is evident in the majority opinion of the *Daubert* court, and reinforced in the *Joiner* and *Kumho* cases. The law's notions of causation, as expressed in many court decisions of the 1980s, and culminating in the *Daubert-Joiner-Kumho* trio, were surely not intended to weaken the authority of science in civil proceedings. Nevertheless they have had important consequences that can be seen as doing just that.

The concept of causation is deeper and more nuanced than it first appears to be. In the simplest case, cause can be directly associated with a single act. We flip a light switch and the light goes on, so we perceive a direct causal relationship between the act of flipping the switch and the light going on. But what if we flip the switch and the light does not come on? The bulb could be burned out, the power to the building may be off, the circuit breaker might be open, or the switch could be defective. All these elements that must be in working order for the light to come on when we flip the switch are part of a causal chain. We focus on the act of flipping the switch because it is the *proximate* cause of the light coming on.

As another example, the development of clinically evident colon cancer is often preceded by the formation of polyps that are benign, then precancerous, and finally cancerous. A series of changes at the cellular level is necessary to the eventual appearance of colon cancer. Environmental factors, such as diet, as well as genetic factors influence the probability that cancer will develop in any individual, and in how rapidly it progresses from one stage to another. Clearly, it makes no sense to speak of a single cause of colon cancer, but rather of a causal chain that may involve many elements. A *sufficient cause*, which is a complete causal mechanism, is defined as a set of minimal conditions and events necessary to produce disease. The end result might be forestalled by interference with any one of the elements along the chain. On the other hand, dietary habits or ingestion of particular toxins as a result of environmental conditions could act to promote colon cancer. If, however, they affect only part of a causal chain, their effects may not be evident for some time. When causation operates over an extended period, assignment of causal roles to the various potential elements in a hypothesized causal chain may be impossible.

Questions of causation are, of course, at the heart of tort litigation. The plaintiff seeks redress for some damage that can be attributed to acts of the defendant. The plaintiffs who had used Bendectin to relieve symptoms of morning sickness sought recovery from Merrell Dow Pharmaceuticals Inc. for birth defects allegedly resulting from the mother's use of the drug. But how could it be established that Bendectin was in fact responsible for any individual birth defect? A direct inference of this kind could be reliably made only if the incidence of birth defects among the cohort of infants born to mothers who had taken Bendectin were far above the incidence of birth defects in the general population of newborns. That, however, was not the case. Birth defects are not uncommon; it has been estimated that there are on the order of 250,000 infants born with birth defects annually in the United States (Schardein 2000). Bendectin was widely used between 1956 and 1983; at some points in that interval, as many as 25 percent of newly pregnant women may have been taking the drug. It would thus be expected that many thousands of infants with birth defects would be born to women who had taken Bendectin, quite independently of the drug. Nonetheless, many women who had given birth to infants with defects, particularly limb-reduction defects, sued for damages in the belief that Bendectin had been the cause.

On the whole, plaintiffs did not fare well in the Bendectin suits. Although a few epidemiological studies were reported to find associations between the drug and one or another type of birth defect, the associations were weak, and confounding factors could not be excluded. In one of the trials, a class action trial on behalf of nearly one thousand infants in Cincinnati (*Richardson v. Richardson-Merrell Inc* 1985), the court admitted evidence presented by the plaintiffs showing causation. The jury, however, found it insufficient to meet the burden of proof on the issue of causation, and returned a verdict for the defendant. In two other notable cases, the plaintiffs' evidence was admitted and the juries found for the plaintiffs. In both of those cases, however, the appeals court or trial judge nullified the jury decision by granting the defendant's motion for a verdict *non obstante veredicto* (notwithstanding the verdict). The courts took the position that the opinions offered by the plaintiffs' expert witnesses were insufficient to demonstrate causation by "a preponderance of the evidence."

It is evident that the *Daubert* court was drawing on a substantial body of legal precedent in its examination of the grounds for admissibility of scientific evidence. The court established a two-pronged test: scientific evidence must be both relevant and reliable. Relevance means that the evidence presented fits the facts of the case. Reliability, the really new component of the test, means that the expert must have employed the scientific method in arriving at her conclusions Among the characteristics to be looked for in evaluating reliability are that the hypotheses motivating the study were properly tested; that the work had been peer-reviewed and published in scientific journals of acceptable quality; that the techniques employed were subject to standards of evaluation and were generally recognized in the relevant scientific community; and that the known or potential rates of error were accounted for (Berger 2005).

On the face of it, the *Daubert* decision placed a higher burden on plaintiffs, by enjoining judges to more actively evaluate prospective testimony, and by opening the door to directed verdicts, once trial begins, if the plaintiff's evidence is found inadmissible. Shortly after the *Daubert* decision, the Supreme Court added further to the plaintiff's burden by ruling in *General Electric Company v. Joiner* (1997) that appellate courts must use an "abuse of discretion" standard in reviewing a lower court's evidentiary ruling. That is, the appellate court could not simply second-guess the trial court; the lower court's ruling had to be manifestly erroneous.

So far, nothing I have said would suggest other than that the courts have been increasingly at pains to enhance the influence of science in the courtroom. The problem, however, is that court rulings since these landmark decisions reveal that judges have shown a penchant for accepting epidemiological evidence to the virtual exclusion of other kinds of evidence that might bear on causation.[5] But epidemiological studies are not the sharp instruments for assessing causation that they are often portrayed as being. There exists no set of necessary and sufficient causal criteria that can unambiguously identify a causal connection on the basis of an epidemiological study. What does emerge is something called a "probability of causation." The nonexpert can be confused by the plethora of technical details that arise in considering the many ways to assess causation and causal inference (Greenland and Robins 2000; Rothman and Greenland 2005). Furthermore, an epidemiological study may be compromised in a variety of ways, including an inappropriate sample population, insufficient sample size, selection bias, confounding factors such as latency, and other contributing factors not identified. There has been a persistent gap between the number of clinical trials conducted and the number for which results are publicly available (Zarin and Tse 2008). Trials are not made available or are incompletely reported for a variety of reasons, including industrial concerns

---

5. Between 1995 and 1999, the percentage of summary judgments resulting in case dismissals increased greatly. The vast majority of these summary judgments were granted in response to defendants' challenges to evidence presented by plaintiffs (Dixon and Gill 2001).

about competitive advantage (Kaiser 2008). There is also an understandable reluctance to make public any post-market observational studies that point to unanticipated adverse effects.

To add to the potential shortcomings of epidemiological studies, a naive reliance on shortcut rules sometimes leads judges to evaluate epidemiological evidence of causation in terms of a risk ratio of 2.0 in exposed populations. This figure can be interpreted as a doubling in the risk of disease, or whatever is being measured, as compared with the general population. In the view of some jurists, a risk ratio of 2.0 or greater is required to demonstrate that is it more likely than not that the disease is attributable to the exposure. There are, however, many logical as well as methodological problems with the unsophisticated application of such a criterion (D. Barnes 2001; Melnick 2005; Cranor 2006, chap. 6).

For all the foregoing reasons, reliance on epidemiological evidence to the relative exclusion of other forms of scientific testimony in tort litigation often works to the disadvantage of the plaintiffs:

> Courts have expressed a preference for the epidemiological study because it relates to human effects. But such studies may not be available when the plaintiff needs to institute an action in order to avoid having the suit barred by the statute of limitation. Or it may be impossible to do an epidemiological study because the product has been taken off the market or because the condition from which the plaintiff suffers is so rare. In addition, such studies are expensive and time-consuming and subject to confounders and biases. Nevertheless, some courts have suggested that plaintiffs cannot win in the absence of a positive epidemiological study, even when the defendant has far more available information and resources to undertake such a study, and even though the scientific community looks to other forms of proof in assessing causation, especially animal studies. (Berger 2005, S61–S62)

There is much evidence to suggest that the *Daubert-Joiner-Kumho* trio of cases has not had the effect of enhancing the authority of science in the courtroom. It has been argued that one of the principal impediments to the appropriate use of science has been an idealization of science, in which science is believed by some jurists to have the capacity to deliver unproblematic judgment on a matter before the court (Caudill and LaRue 2006). When it fails to do that—because the scientific results are less than certain, because the issues are not entirely clear, or because social factors cloud the issues—the scientific evidence presented may be rejected, even though it may have probative value. In the survey of a large sample of state court judges referred to earlier, when the judges were asked to define or explain each of the four *Daubert* factors, only a small percentage could demonstrate a working understanding of falsifiability or error rate (Gatowski et al. 2001). Although I am not aware of any studies of judges' understanding of causation and how it relates to the burden of proof, there is ample evidence that many judges are not able to relate scientific practices of drawing inferences from diverse sources of information to the preponderance of evidence standard that should apply in their courtroom (Cranor 2005). For example, there is a strong tendency to exclude animal studies. Yet, with respect to carcinogenicity, for example, it is widely recognized as prudent to treat substances in which there is sufficient evidence of carcinogenicity in animals as though they present a carcinogenic risk in humans. Virtually all known human carcinogens that have been adequately tested have been found to produce significant carcinogenic effects in one or more animal models (Weinstein 1991). As an example, the chemical 1,3-butadiene was found to be a potent multi-organ carcinogen in animals at dosage levels considerably below the Occupational Safety and Health Administration exposure limit for humans. Later epidemiological studies confirmed the carcinogenicity of the substance in humans, and the occupational exposure limit was revised downward, from one thousand parts per million to one part per million.

Admittedly, animal studies do not always reveal a close correspondence between toxicity in any particular animal model and in humans. For example, although the chemical dioxin is a potent carcinogen in certain animals, evidence based on a large number of studies suggests that it is much less toxic in humans (Gough 1993). Differences of this degree may happen because the animal studies are conducted at higher dosage levels out of practical considerations of time and scope of the study, and because a meaningful comparison to human populations requires a large sample data base that may not be available. Further, there may not be sufficient time for human studies to be completed, given that latency effects may operate in carcinogenesis. In addition, no single animal species furnishes a best model for all classes of potentially toxic substances (Schardein 2000, 28). Despite these limitations, animal studies have value precisely because they can be conducted and evaluated when appropriate studies of human populations are not feasible. A National Academies commission report on evaluating chemical exposures for their reproductive and developmental toxicity refers extensively to the utility of both animal and in vitro studies as important indicators of potential for toxicity in humans (NRC 2001). The issue then should not turn on whether there is a proven correlation between any individual animal study and a human study; rather, it should be whether there is a reasonable expectation that the results of the animal or in vitro studies will be applicable to humans to a significant degree in specific cases (Green 1992, 654–56; Cranor 2006).

Forms of evidence based on laboratory experiments are also capable of providing additional grounds for assessing whether a substance could produce particular biological effects in humans. In vitro studies involve testing the effect of the substance on biological substrates such as cells, bacterial cultures, organs, or embryos. Broad correlations have been established between the effects of substances on such substrates and their effects in humans. As noted, however, exceptional cases arise, so such methods fall short of establishing causation. Nevertheless, they can, in company with other evidence, form part of a package of evidence that bears on a question of causation.

In recent years, advances in drug screening methods, coupled with the advent of powerful, computer-based modeling programs, have greatly enhanced scientists' capacity to design drugs for specific pharmacological effects, and to reduce undesirable side effects. Pharmaceutical companies spend millions of dollars each year on such research, which has become an integral part of the drug discovery process. These methods are often coupled with laboratory studies. As an example, scientists at Pfizer have developed a tool called "biological activity spectra," derived from in vitro measurements of the binding of the drug to a wide range of proteins (Fliri et al. 2005). They applied their method to an analysis of more than one thousand prescription drugs for which adverse effects were identified on the labels, and were able to show strong correlations of effects with chemical structure. In addition to the promise that such advances offer for improving drug design, these methods are emblematic of increasingly detailed and reliable correlations between structures and in vitro properties, on the one hand, with clinical properties, on the other. Studies of systems at the molecular and cellular level, directed toward observing how chemicals interrupt cellular activities, will in time provide detailed understandings of which kinds of chemical structures elicit particular effects within cells (Schreiber 2005). As such methods are further developed, they will be powerful tools in assessing whether a particular substance might be an environmental hazard or an effective pharmaceutical, as well as whether it might be capable of producing a toxic effect. Furthermore, it is becoming increasingly feasible to associate drug efficacy and potential toxic effects in some cases with particular genetic markers, thus providing a means of identifying individuals likely to be at risk for adverse effects.[6] Given that the potential for adverse reactions to a drug may vary with genetically controlled factors, epidemiological studies alone can hardly be expected to provide an adequate basis for judging a drug's effect on a particular individual.

I have merely hinted at what is in fact a vast research effort on many fronts, which has transformed our understanding of how chemical substances, whether as drugs or environmental agents, operate in human biology; how individual differences in genetic makeup can lead to different outcomes with a given substance; and a host of other topics that might arise in the course of litigation. But how is this developing science going to find expression in the courtroom? Some of it might turn up in the course of discovery, as when a pharmaceutical company has carried out extensive in vitro studies that include a drug that is the subject of a tort suit. Some of it might be brought by plaintiffs in an effort to establish causation. In general, however, there is a lot of relevant science, already accomplished or in progress, that will not find the role in legal decision making that it is capable of playing (Schulz 2006; Cranor 2006, chap. 7).

Evidence derived from such increasingly sophisticated means of establishing relationships between chemical structures and biological effects does not, of course, prove causation. Nevertheless, it may prove sufficiently robust to support a preponderance of the evidence argument for or against. It could prove to be more persuasive than the sometimes flawed or limited epidemiological studies that many judges seem to regard as the most reliable measures of a causal relationship. To be considered at all, however, it must be admitted into evidence. Of course it not the judge's job to seek out new lines of scientific evidence. But when new forms of scientific evidence are brought forth, as they surely will be, judges will be faced with the continuing challenge of evaluating them in light of the *Daubert* criteria. To judge from judicial decisions of the past couple of decades, their evaluations are likely to be very conservative. When an argument for a causal connection is based on a package of related items, courts tend to evaluate each piece separately from the others, instead of considering the weight of all the evidence together in support of an explanatory conclusion.

---

6. The literature on this subject is vast. Two representative examples are Weinshiboum 2003 and J. Wilson et al. 2001.

## A SUMMARY AND SOME FURTHER THOUGHTS

Our examination of how science contributes to the administration of justice in criminal and civil cases, though incomplete in many respects, reveals several important facets that limit the authority of science in the courts.

First, the appearance of scientific consensus is often lacking. We saw earlier that consensus within the scientific community on scientific questions at issue is one of the pillars on which authority is constructed. The adversarial nature of courtroom proceedings that involve litigation, however, virtually ensures that scientific claims will be challenged vigorously. It is thus understandable that controversies regarding certain underlying scientific issues in the field of DNA typing had the effect of weakening the probative value of DNA evidence in the courts. The scientific community had to work to establish an image of consensus in this area of science in order to regain authority in legal proceedings that turned on DNA evidence. A deeper understanding of the nature of scientific inquiry, including the fact that scientific claims are *always* subject to challenge, would help both jurists and juries avoid unproductive idealizations of science that often limit science's legitimate authority to participate in legal proceedings (Caudill and LaRue 2006).

Second, because the courtroom is an adversarial arena, scientists who testify are generally assumed not to be disinterested parties. Those who testify for the plaintiff in a tort case are assumed by the defense to be offering biased, incomplete evidence and opinion regarding the matter at hand. The plaintiffs take scientists testifying for the defendants to be similarly nonobjective in their testimony. Biases of course arise in the ordinary give-and-take of scientific discourse, but the ground rules of science are greatly different. It is deemed unethical to deliberately conceal results that might weaken one's position with respect to a contested hypothesis or claim. In legal proceedings, it might take discovery procedures or rigorous cross-examination to uncover unfavorable data or observations. Thus, one of the mainstays of science's claims to authority—that the scientist is disinterested, objective, and fully forthcoming in evaluating and presenting evidence—is weakened. Further, many with credentials that allow them to pass muster in the courtroom as scientific experts are more or less career expert witnesses in tort cases, often testifying to matters with which they have little direct experience as scientists.

In principle, scientific authority might be enhanced through the use of court-appointed science witnesses, counsels, or panels (Hooper, Cecil, and

Willgang 2001). In practice, however, judges have shown little enthusiasm for such appointments (Saks and Faigman 2005; Haack 2005). In practice, the ideal of epistemically sound, unbiased scientific expertise can fall short of the mark in several respects (Mnookin 2008). When experts have been appointed, their operations have often proved cumbersome. Further, in the adversarial contexts commonly encountered, they almost always incur charges of bias or conflicts of interest. The search should be continued for alternative mechanisms for providing courts with quality scientific expertise that is, insofar as possible, free of biases specific to the case at hand.

Third, judges generally lack sufficient understanding of science to fulfill their gatekeeper role. Though judges typically lack sufficient knowledge of science and of scientific methods, they are often required to rule on the admissibility of scientific evidence. The amount and quality of science that makes it into the courtroom for presentation to the triers of fact, the jurors, is limited by the biases and scientific naivety of the judge, acting as gatekeeper. It is the judge's job to serve the cause of justice by excluding testimony that does not promise to illuminate the issues and assist the triers of fact. In the past two decades, a few prestigious judges have summarily excluded categories of evidence on the grounds that they do not demonstrate or prove a causal relationship. Their decisions in tort cases involving alleged adverse effects from drugs or chemicals in the environment have strongly influenced other jurists. The result appears to have been prejudicial to the presentation of much evidence that scientists would regard as valid in drawing inferences. Surveys have shown that judges are not generally able to evaluate epidemiological studies in terms of such factors as sample size, error rate, and the like. Thus, such studies may be admitted in evidence even though they do not meet the *Daubert* criteria. Further, because they do not in themselves constitute scientific evidence of diseases or adverse reactions, epidemiological studies can provide at most only an indirect suggestion of causation. By contrast, biomedical research, including clinical studies of patients, animal studies, in vitro studies, and chemical biological methods can yield evidence that permits formulation of hypotheses or theories that directly address questions of causation. They do not typically *prove* causation, which is a very high standard to achieve. In the courtroom a "preponderance of the evidence" is, or should be, the standard; the fact in controversy must be deemed more likely than not. Furthermore, it is not the judges' prerogative to decide that issue in advance, but only to ensure that evidence that can usefully contribute to making the decision is available to the triers of fact.

Finally, the courts are slow to incorporate scientific advances. I have noted above that many fields of science that can in principle afford evidence regarding causation are advancing very rapidly. Most advances in fields such as molecular biology, genetics, chemical biology, and environmental science find rapid publication, so the issue of proper peer review does not constitute a barrier to their admission into the courtroom. But mechanisms for identifying key scientific studies that could be of relevance in particular cases do not always exist. Law enforcement agencies dealing with criminal investigations are oriented toward exploring the interfaces with various areas of science, so new technologies can be adapted to forensic purposes fairly quickly. In tort law, however, plaintiffs seldom have the expertise to know where to look for the kinds of scientific evidence that could support their claims. Further, most scientists are not eager to take time and energy away from research to apply some of their results and expertise to legal cases. In product liability suits, the defendants may have conducted in vitro or computer-based modeling studies on the products involved in the suit, but the results of such studies require expert evaluation as to their relevance and reliability. It may not be easy for plaintiffs to obtain the expertise for such evaluations.

This brief examination of science in the courtroom has touched on only a few areas of law in which the authority of science comes into play. Statutory and regulatory law, which we will consider in chapters dealing with government and religion, are yet other arenas in which the authority of science is exercised and challenged. We will see that the challenges to science's authority there are often of a different kind than those that obtain in criminal and civil law.

# SIX

## SCIENCE AND RELIGION

The epistemic and moral authority of science in American society are more vigorously and publicly challenged through its conflicts with religion than with any other sector of society. Religion and science are frequently seen to be in dramatic opposition. Consider, for example, the contentious matter of teaching evolution in public schools; restrictions on the uses of pluripotent ("embryonic") stem cells for biomedical research; contested policies regarding the use of government funds to provide contraceptives or birth control information; and attempts to limit funding for AIDS research. In all these instances, arguments based on science stand on one side and those based on one or another religious outlook stand on the other. What is it about the contrasting natures of science and religion that make such conflicts so frequent? Are the premises of the two domains somehow inherently irreconcilable? Some insist that this is the case: Richard Dawkins, E. O. Wilson, and Lewis Wolpert, for example, are avowedly atheist, or secular humanist, in their outlooks; others, including scientists such as Francis S. Collins, find it possible to reconcile their religious beliefs with their scientific approaches.

### WHAT IS RELIGION?

Before we can usefully look at specific instances in which science and religion seem to be in conflict, we must clarify what we mean by religion and what it means to have a religious outlook. We must also examine the comparative natures of religious and scientific authorities. Max Weber began *The Sociology of Religion* (1922) by saying that although religion is largely manifested as a collective, social phenomenon, it ultimately reduces to individual subjective experience, which varies greatly from one person to another in content and intensity:

> It is not possible to define religion, to say what it "is," at the start of a presentation such as this. Definition can be attempted, if at all, only at the conclusion of the study. The "essence" of religion is not even our concern, as we make it our task to study the conditions and effects of a particular type of social action. The external courses of religious behavior are so diverse that an understanding of this behavior can only be achieved from the viewpoint of the subjective experiences, notion, and purposes of the individuals concerned—in short, from the viewpoint of the religious behavior's "meaning." (1)

We will be concerned primarily with religious belief in a supernatural order, where, according to Rodney Stark, "supernatural refers to *forces or entities* (conscious or not) that are *beyond or outside nature* and which can suspend, alter, or ignore physical forces. Gods are a particular form of the supernatural consisting of *conscious supernatural beings*" (2003, 4; emphases in the original). Further, our attention will be directed primarily toward what Stark refers to as "Godly Religions," because these give rise to the most evident conflicts of authority with science. Stark notes that most people prefer a godly religion, which rests on revelations and admits of communications from the gods (and perhaps to the gods as well):

> Why do most people prefer a Godly religion? Because Gods are the only plausible sources of many things people desire intensely. It must be recognized that these desires are not limited to tangibles. Very often it is rewards of the spirit that people seek from the Gods: meaning, dignity, hope, and inspiration. Even so, the most basic aspect of religious activity consists of exchange relations between humans and Gods: people ask of the Gods and make sacrifices to them. Indeed, it is believed that Gods, unlike unconscious essences, set the terms for such exchanges and communicate them to humans. . . . Godly religions rest upon *revelations*, on *communications believed to come from the Gods*. (5)

Stark writes generally here in terms of gods, but his focus and ours is on monotheistic religions, and on Christianity in particular.

As Weber recognized, religion is ultimately a private and personal part of individual lives, whereas much of our concern here is with religion manifested primarily as a social or collective phenomenon. The private and social are connected because, for most people, religious life in terms of salvation, conscience, and relationship with the divine is sustained by communal agreements on what constitutes accepted belief, and on appropriate rituals, observances, and taboos. Mutual reassurance is essential to forming and sustaining such beliefs, which take form over time, both in the individual and in the community as a whole. Émile Durkheim identified the essentially social character of religion: "A religion is a unified system of beliefs and practices relative to sacred things, that is to say, things set apart and forbidden—beliefs and practices which unite into one single moral community called a Church, all those who adhere to them" (1912, 44). Religion exercises a powerful cultural authority that exists alongside its traditional authority. Looked at empirically, this is true whether one believes with Durkheim or Weber that religion is social all the way down, or conversely that it is grounded in supernatural reality (Stark and Bainbridge 1987).

Granting that religion of the sort that interests us here is importantly social in nature, it is also true that it is also highly pluralistic. One of the reasons for this is that people seem to be spread over a wide spectrum in terms of their religious tastes. Max Weber noted that in every religious organization people differ in the extent of their religious commitments. He spoke of those in any religious society who are most intense about their religiosity as "religious virtuosi" (1912, 162): monks, hermits, ascetics of various kinds. According to Stark, "Some people are content with a religion that, although it promises less, also requires less. Others want more from their religion and are willing to do more to get it. . . . Thus, in any society where diversity is not suppressed by force, the religious spectrum will include a full range of religious organizations, from some that demand little and are in a low state of tension with their surroundings to some that offer very high intensity faith" (2003, 17–18). New sects, particularly ones that promise a more intense religious outlook, are formed when people are not satisfied with the state of the church to which they belong. Sects change over time; internal challenges to the sect's belief system arise, societal changes engender diverging outlooks, and perhaps geographical factors play a role. Whatever the driving force, in the course of time new sects form, even as others die out. The result is that religion in American society is composed of a diverse array of communities with sometimes disparate beliefs and practices. There is no denying, however, that evangelical, conservative Christianity has been a peculiarly American phenomenon that has played a powerful role in shaping the nation's cultural landscape (Hofstadter 1963, chap. 3; Jacoby 2008, chap. 8).

## THE CONTRASTING AUTHORITIES OF SCIENCE AND RELIGION

Stephen Jay Gould was fond of talking about science and religion as "non-overlapping magisteria," by which he meant that each has a legitimate domain of teaching authority that does not interfere with the other's. He first proposed this view in an article dealing with the Catholic Church's papal pronouncements regarding evolution (Gould 1997). Though Gould was a professed agnostic, his conviction that there is no necessary irreconcilability between science and religion is shared by many scientists, whose religious beliefs vary widely (Polinghorne 1998; Collins 2006; Roughgarden 2006; Gingerich 2006). On the other hand, many scientists and philosophers hold that science and religion are incompatible in their outlooks and claims (Weinberg 1994; E. O. Wilson, 1999, 2006; Raymo 1998; Haack 2003, chap. 10; Dennett 2006; Harris 2006; Dawkins 2006; Wolpert 2007). Why do intelligent, sophisticated, and accomplished people, all of whom presumably have access to the same general store of information and cultural outlooks, come to such different conclusions on this question? At least part of the answer can be framed in terms of the contrasting kinds of authority wielded by science and religion.

At the simplest level, religious and scientific authorities are of different kinds. In terms of the classification outlined in chapter 1, scientific authority has *rational-legal* social origins and is primarily expert in nature. The methods of science and its body of accumulated observations and theoretical interpretations form the basis on which scientists speak with expert knowledge of how things are in the physical world. Religious authority, on the other hand, is of a *traditional*, institutional nature. It rests on a body of revelation that the individual accepts on faith. Revelation is concerned primarily with supernatural matters, though religious acts and ideas are frequently conveyed via agency in the physical world. Conflicts between science and religion arise when one makes claims that contradict accepted beliefs within the domain of the other.

The purported conflicts between science and religion that receive widespread attention would seem to be cases of institutional conflicts. For example, a religious denomination or other organized religious entity might make a claim regarding the age of the earth that is in conflict with scientific evidence consensual within the scientific community. To understand how such conflicting positions arise and are given expression at an institutional level, however, we must look beneath the larger-scale appearances, to the level of individual perceptions and beliefs. That assessment leads us to ask how these two very different domains exercise *cultural* authority.

Science is a powerful force in society, even in the face of widespread public ignorance of basic scientific understanding of how the world works, of the methods of science, of what constitutes a valid scientific claim—in short, of what science is really all about. This largely unknown domain of human activity is accorded respect, albeit sometimes mixed with fear and distrust, because science has been responsible, through the agency of technological application, for a great deal of what constitutes the modern world. Science is widely appreciated as being effective; in this sense, it exercises a pervasive cultural authority. The man or woman in the white lab coat appears frequently in TV ads to assure us that a product, whether a household cleaner, a sleep medication, or a new car design, passes muster scientifically. Studies generally accepted as "scientific" guide our thoughts regarding the dietary health effects of olive oil, the influence of computer-based games on cognitive development of children, the long-term effects of the Exxon Valdez oil spill on marine life in Prince William Sound (National Oceanic and Atmospheric Administration 2008), and a great deal more. Though people enjoy material benefits deriving from science and technology, they are not thereby imbued with the attitudes of curiosity and skepticism that are the hallmarks of a scientific outlook. The physical principles that underlie the operation of a CD player or an iPod are neither apparent nor of interest to nearly all the millions who use such devices.

Thus, it would be a mistake to suppose that science exercises a significant direct influence on peoples' outlooks on life. Certainly, there are indirect effects: improvements in health and living standards generally have made possible increased access to education, to longer life expectancy, to greater ease of communication of family members with one another, and so on. All of these benefits have led to changed outlooks on life. The sense that science and technology lie behind so much that contributes positively to the quality of modern life, however, lies pretty much in the background, largely taken for granted. Only infrequently are we brought up short with a realization that modern science and technology are vital to our lives: a CAT scan reveals a hitherto-undetected aortic aneurism; a GPS device in a car directs emergency crews to the scene of an accident. Science and technology thus exercise a broad but somewhat diffuse cultural authority. Science is understood as a creative force able to make things happen. Because it is seen as a source of power, it commands respect and attention, even on the part of people who generally do not understand how it works.

Religion and science are alike in exercising authority over the vast majority of people in the absence of their firsthand knowledge of the claims made. The hallmark of what Rodney Stark refers to as godly religions is belief in a supernatural being who is the creator of the world and responsible for its continued existence. In its weakest form, the creator is not supposed to be an active participant in the workings of the world. The "big bang" creator is thought to have made the universe at the instant of its creation, and allowed it to evolve in an unattended, free-running mode ever since. At the opposite end of the spectrum of beliefs are those who believe that God not only created the world, but that he is continuously attentive to its workings and actively influences events. Most commonly (if we confine ourselves to Christians), such people look to the Bible as the primary source of religious truth. Some insist on a literal interpretation of the biblical text. They may insist that the world was created in seven days, that the earth is no more than several thousand years old, that there was a great flood, that Noah built an ark that held all the animal species that survived the flood, and that all the miracles described in the Bible actually occurred. Wherever along this spectrum an individual's beliefs may lie, none of them is the product of direct experience, nor are they verifiable.

Many people, including scientists, hold that it is not inconsistent to accept the dogma of one or another sect and at the same time adopt the rationalistic, skeptical outlook of a scientist toward the physical world. Others argue that with respect to much of what most Christian sects proclaim, such a position is untenable. How can a question of this kind be settled? A good deal turns on what level of philosophical rigor one applies in testing beliefs, both scientific and religious. Most people, of course, do not give much attention to philosophical niceties. They are not inclined to plumb the depths of their beliefs looking for intellectual shoals, for inconsistencies in logical structures, or even for that matter looking for logical structures at all. Rather, their belief systems, especially in the case of religion, are the products of formations that might have begun early in life. They are shaped by various family and societal influences, as well as strongly emotional personal experiences. In such manner religious beliefs may be formed in the absence of a deep assessment of whether they are consistent with the canons of science. Similarly, the development of a scientific outlook in the course of an apprenticeship in science normally occurs without specific references to potential conflicts with religious beliefs. Individuals are left to resolve for themselves whatever conflicts they discern.

As an illustration of the dichotomous beliefs and attitudes that can arise at the intersection of science and religion, consider the survey conducted in 1914 by James H. Leuba, a Bryn Mawr College psychologist. Leuba wanted to know the views of scientists on what he considered two important Christian religious concepts: a god who is attentive and responsive to humans, and the existence of an afterlife. His survey consisted of the questions shown in Figure 1.

Leuba sent his questionnaire to one thousand persons selected at random from the 1910 edition of *American Men of Science*, and received about

---

**A. Concerning the Belief in God**

1. I believe in a God in intellectual and affective communication with humankind, i.e., a God to whom one may pray in expectation of receiving an answer. By "answer" I mean more than the subjective, psychological effect of prayer.

2. I do not believe in a God as defined above.

3. I have no definite belief regarding this question.

**B. Concerning the Belief in Personal Immortality**

   *i.e., the belief in continuation of the person after death in another world*

1. I believe in:
   a. personal immortality for all people
   b. conditional immortality, i.e., for those who have reached a certain state of development.

2. I believe in neither conditional nor unconditional immortality of the person.

3. I have no definite belief regarding this question.

4. Although I cannot believe in personal immortality I desire it:
   a. intensely
   b. moderately
   c. not at all.

---

**Figure 1** Questions from Leuba's scientist survey

a 70 percent rate of response. The results are summarized in Table 1 under the 1916 heading (Leuba 1916). Leuba's survey was repeated in 1996 by Edward J. Larson and Larry Witham; they randomly chose one thousand names from the then-current edition of *American Men and Women of Science*, and received about a 60 percent rate of response (Larson and Witham 1997). Their results are listed in Table 1 under the 1996 heading. Remarkably, the proportion of those surveyed who believe in a personal god remained relatively constant at about 40 percent, though the two surveys were eighty years apart. The high percentage of scientists in 1914 who disbelieved in a personal god or were agnostic about the matter was something of a scandal for William Jennings Bryan and conservative Christians; they feared that academic disbelievers were leading college youth away from a religious outlook. Conversely, in 1996, the percentage of scientists surveyed who professed a belief might have come as a pleasant surprise to conservative Christians, who are inclined to view scientists as a generally godless lot.

The questions posed in these surveys were criticized by many of those who received the survey or examined the results. The small size of the query set left no room for affirming a belief that God exists but does not necessarily relate to individuals in a personal way. Nevertheless, responses to the questions asked seem quite unequivocally to show that some 40 percent of those surveyed believed that God can be counted on to intervene in the workings of the physical world. Over those eighty years, however, the fraction of those surveyed who held a belief in personal immortality declined precipitously.

**Table 1** Summary of Leuba's scientist survey*

|  | 1916 | 1996 |
|---|---|---|
| *Question A (belief in a personal god):* | | |
| Response 1 | 41.8% | 39.3% |
| Response 2 | 41.5% | 45.3% |
| Response 3 | 16.7% | 14.5% |
| *Question B (belief in personal immortality):* | | |
| Response 1 | 50.6% | 38.0% |
| Response 2 | about 20% | 46.9% |
| Response 3 | about 30% | 15.0% |
| *Desire for personal immortality:* | | |
| Intense | 34% | 9.9% |
| Moderate | 39% | 25.9% |
| Not at all | 27% | 64.2% |

*The percentages given in the table are those listed by Larson and Witham.

One might ask about the extent to which Leuba's survey population could truly be counted as practicing research scientists. Conscious of this problem, Leuba also surveyed a set of scientists who had been designated by the editors of *American Men of Science* as especially noteworthy, the scientific stars of their time. In that subset, he found distinctly lower levels of belief in a personal god or in the prospects for immortality. In a comparable effort, Larson and Witham in 1998 surveyed all the members of the NAS in the core biological and physical sciences, and found that more than 90 percent of the respondents expressed disbelief (Larson and Witham 1999). Larson and Witham credit Rodney Stark with the view that there exist social pressures in science against holding religious views, suggesting that in the course of their development, religious beliefs are somehow wrung out of young aspiring scientists. This unlikely hypothesis seems to me to betray a lack of familiarity with the culture of the typical physical or biological science laboratory, where there is little time or inclination to indulge in religious speculation of any kind. It seems more likely that as young scientists become imbued with the rational methods, skeptical outlook, and accomplishments of science, the idea of a personal god or of immortality becomes increasingly incredible.

What can we make of the Leuba/Larson-Witham survey results? Clearly, for the most eminent and successful scientists, scientific authority dominates over religious authority. If one does not believe in the existence of a personal god, or in personal immortality, there is not much ground on which religious authority could be exercised. In the more general survey, a substantial fraction of those surveyed—people whose credentials as scientists are likely mixed, but good enough to gain them entry into *Men and Women of American Science*—assert a belief in a personal god who can be expected on some occasions to act in the world to respond to prayer. Is such belief inconsistent with the tenets of modern science? Clearly, many eminent scientists, Charles H. Townes and Francis Collins among them, think not. Townes, a Nobel laureate in physics (1964) and winner of the 2005 Templeton Prize,[1] has long been known for his view that science and religion are alike in seeking truth that transcends human understanding, and in seeking meaning in the universe. In the introduction to his book *The Language of God: A Scientist Presents Evidence for Belief* (2006),

---

1. Currently valued at about two million dollars, and awarded for making "an exceptional contribution to affirming life's spiritual dimension, whether through insight, discovery, or practical works." The foundation was established by John Templeton in 1972 (John Templeton Foundation 2008).

Collins writes: "In this modern era of cosmology, evolution and the human genome, is there still a possibility of a richly satisfying harmony between the scientific and spiritual worldviews? I answer with a resounding yes! In my view, there is no conflict in being a rigorous scientist and a person who believes in a God who takes a personal interest in each one of us. Science's domain is to explore nature. God's domain is in the spiritual world, a realm not possible to explore with the tools and language of science" (6).

But is there not an inconsistency in believing the two realms of science and religion to be somehow separate, and yet rejecting—as Collins does—Gould's conception of separate, "non-overlapping magisteria"? If God works in the world at all, if he contravenes the laws of physics and chemistry so as to alter the course of events, the line between the spiritual and natural has been crossed. Scientific naturalism is the doctrine that the world's causal processes are not subject to interruption. It rejects the possibility of a direct supernatural effect on the web of cause-effect relationships. A scientific naturalist would be obliged to reject the idea that there exists a personal god who chooses to respond to prayer by altering things in the world. The survey results would seem to indicate that a fairly large fraction of those counted as scientists are, at a minimum, not scientific naturalists.

Philosophers have struggled with the questions and contradictions wrapped up in the idea of scientific naturalism. We will not make much progress with our project of analyzing authority if we sail into those murky waters. It has been argued, not necessarily successfully, that a thoroughgoing materialistic naturalism that denies the existence of God is incapable of producing a coherent philosophical account of how we can know much of anything (Griffin 2002). Versions of scientific naturalism that maintain the existence of a god that refrains from fiddling with his creation from time to time, however, can exist alongside science. At the opposite pole, belief in the literal truth of the Bible is inconsistent with science, not only in terms of its variance with scientific evidences about the world, but also in terms of the intellectual outlook that rejects the authority of science to testify at all on matters of religious belief.

As we examine specific examples of apparent conflicts between religion and science that have gained wide public attention, we must keep in mind that these contests, if they can be called that, are multilayered. The issues are argued in public forums—the courtroom, school board meetings, legislative sessions—where the conflict between religion and science comes in various guises. At heart, however, the arguments forwarded in those forums are reflective of the religious sensibilities, or lack thereof, of the citizenry at large. When we look for where religion and science attempt to exercise authority, we must look not only at the visible public proceedings, but also at the channels through which public opinions are formed.

## DARWINISM AND RELIGION

Before Charles Darwin published his *On the Origin of Species*, he had studied works such as Charles Lyell's *Principles of Geology*, published in 1830. Lyell's book is very important because it changed the way people viewed Earth's history. Lyell argued that the planet's crust was formed by numerous small changes occurring over vast periods of time, all according to known natural laws. His "uniformitarian" proposal was that the forces molding Earth today have operated continuously throughout its history. We can read the past history of the planet by examining the current geological record. Lyell's work, along with that of others more concerned with biological evidences, contributed to the gradual accommodation in Protestant England to the idea that the biblical account of creation was not to be taken literally. It became acceptable to think of revelation as not teaching facts about the world; the biblical writers took the facts of nature as they appear to ordinary people, and wove metaphorical stories to impart spiritual and moral lessons. Darwin himself seems to have progressed in the course of his studies from traditional Christian beliefs, including a literal interpretation of the Bible, to a general deism. In his personal beliefs he became an ardent evolutionist—that is, one who believes in the general idea of a progressivist change in the natural order through some kind of evolutionary process (Ruse 2005, chap. 6; Gopnik 2006).[2]

The particular question of human origins, addressed by Darwin in *The Descent of Man*, published in 1871, occasioned a great backlash, much of it from religious quarters. In more general terms, however, the idea of evolution came to be widely accepted, particularly among the more highly educated. There was much interest in the huge prehistoric fossils that were found in abundance as the American West was explored, and museums such as the American Museum of Natural History in New York became classrooms for instructing the public in the tenets of evolutionism. In the decades following the Civil War, however, the character of Christian

2. Darwin, however, did not participate in the social Darwinist movement created by Herbert Spencer, which became a progressivist secular religion.

religious practices and beliefs in the United States changed dramatically, and in ways that bore on the status and authority of science (Ruse 2005, chap. 8). Michael Ruse has described how attitudes and justifications for slavery, the effects of the Civil War on the South, the growth of evangelical Christian sects (notably the Baptists and Methodists), and a general absence of a fixed social order in the course of the nation's westward expansion combined to foster the popularity of religious sects that emphasized the authority of the Bible over the interpretive pronouncements of more formal groups: "The big growth occurred in the South, the mid-West, and the far West, not in New England. In the South, for instance, almost 40 percent of the Christian population was Methodist, with about that many Baptists. Episcopalians made up a mere 5 percent. By 1860 about a third of the nation's citizens were actively connected with churches, nearly all of which were Protestant and 85 percent of which were evangelical. The seating capacity of these churches suggests that as much as another third of the population attended church occasionally" (148).

The religiously conservative sects placed great emphasis on individual efforts to attain a state of spiritual purity. Religious motivations for progressive action in the social sphere gave way to a focus on attaining a righteous personal relationship with God, and in helping others in that same effort. The Bible was taken to be the inerrant word of God in all its elements, commanding faith and obedience: "Among the various denominations, conceptions of proper interpretations ranged from those that saw a vital role for scholarship and rational expertise down through a range of increasing enthusiasm and anti-intellectualism to the point at which every individual could reach for *his* Bible and reject the voice of scholarship. After the advent of the higher criticism, the validity of this Biblical individualism became a matter of life or death for fundamentalists" (Hofstadter 1963, 57). In the early part of the twentieth century, a series of twelve pamphlets expounding and defending conservative theology, entitled *The Fundamentals: A Testimony*, was published and broadly distributed (Dickson, Meyer, and Torrey 1910–15). According to Edward B. Davis, the term *fundamentalism* as applied to Protestant theology has its origins in a 1920 definition by Curtis Laws, the editor of a national Baptist weekly. The fundamentalists were to do battle "in defense of certain traditional Christian beliefs against the efforts of liberal Protestants to make those beliefs more consistent with secular thought and culture. As this definition suggests, fundamentalism is best understood as an attitude—a militant rejection of modernity—rather than as a specific set of doctrines" (E. Davis 2005).

Against this outlook were those who espoused a liberal Protestant theology, including many scientists. Evolution came into the picture because
evolutionism, which pushed an agenda of improving human society, was
a strong element in their overall framework. The social Darwinism of Herbert Spencer—he of "survival of the fittest" fame—had become a prominent
feature of the progressivist program, along with the precepts of eugenics.
Many intellectuals saw eugenics as a vehicle for applying scientific knowledge to social problems. To the conservative Christians, all of these liberal
currents, including evolution and eugenics, were an assault on their fundamentalist beliefs. One of their most vigorous and visible spokespersons
was William Jennings Bryan. Although he was strongly progressive in his
social and political outlook, Bryan was also a conservative Christian, and he
saw evolution as providing a rationale for regressive social policies.

The 1925 trial of John Scopes in Tennessee for violating state law, which
forbade the teaching of evolution in public schools, has been the fountainhead of a generous stream of historical accounts (Larson 1997). It was
raised to the proportions of a national spectacle because of the reputations
of the two leading legal figures, William Jennings Bryan for the state, and
Clarence Darrow for the defense. H. L. Mencken's reports from the Rhea
County Courthouse to the *Baltimore Sun*, along with those of many other
reporters to their papers, put the trial on the front pages of newspapers all
over the country. Scopes was convicted of violating the Tennessee law, and
fined one hundred dollars. His conviction was later overturned on a tech-
nicality, thus frustrating his intention to carry the case forward by appeal,
challenging the constitutionality of the law. While the prosecution and the
state of Tennessee were the butt of much scathing humor, many ordinary
people of conservative inclination felt that the Scopes trial came out as it
should have. The antievolution law survived on the books in Tennessee
until 1967, just one year before the United States Supreme Court ruled in
*Epperson v. Arkansas* that the similarly formulated Arkansas law was in violation of the establishment clause of the Constitution. Nor did the Scopes
trial diminish the enthusiasm of religious conservatives for legislation out-
lawing the teaching of evolution; in 1927 antievolution bills were introduced in several states.

In the years leading up to the *Epperson* decision, evolution was not a
prominent feature of public school textbooks. Textbook authors and their
publishers realized that according any prominence to the subject could
jeopardize adoption of a biology book. But things changed in the 1960s; the
Russian launch in 1957 of *Sputnik*, the world's first artificial satellite, came

as a shock to the American public. The threat of Russian technical superiority galvanized steps to improve science education in the public schools, and many curricula came in for modernization. A renewed emphasis on science education, including a revitalization of the biology curriculum, brought evolution back into the classroom and rekindled the long-standing opposition of conservative Christians (Kitcher 1982, 1–6). In the wake of the *Epperson* decision, however, it was no longer feasible to simply prohibit the teaching of evolution.

"Creationists"—those who rejected the evolutionary account of how the world came to be in favor of a Bible-based account—needed to explain away a very extensive body of scientific evidence. In the first half of the twentieth century, various alternative explanations were put forth. For example, one George McCready Price, a Canadian Seventh Day Adventist, sticking to his sect's dogma that the world was created in six literal days, produced a story of how the great flood, which was survived by Noah and his family alone, led to the earth's observed geology. Price's quite incredible rationalizations were more or less reiterated in 1961 by John C. Whitcomb, a theologian, and Henry M. Morris, a hydraulic engineer, in their book *The Genesis Flood: The Biblical Record and Its Scientific Implications*. Morris became a leading spokesman for creationism, and *The Genesis Flood* became a central element of *scientific* creationism, or creation science (Kitcher 1982; Ruse 2005, 245).

The strategy of the creationist movement became one of vigorous criticism of the evidence for evolution (Gish 1979), coupled with insistence that creation science offered a scientifically sound alternative explanation of the evidence adduced to support evolution. Efforts were made in several states to introduce bills that would mandate the teaching of creationism in the public schools as a valid scientific alternative to evolution. In 1981 a bill requiring a "balanced treatment to creation-science and evolution-science" was passed by the legislative houses in Arkansas and signed into law by the governor. The bill was successfully challenged in federal court on the grounds that it violated the intent of the establishment of religion clause of the First Amendment of the Constitution. U.S. district court judge William R. Overton found that the Arkansas statute did not have a secular purpose, noting the use of language peculiar to creationist literature, which emphasized the origins of life as an aspect of the theory of evolution. The scientific community does not consider the question of life's origins an integral part of evolutionary theory, which assumes the existence of life and is directed to an explanation of how it evolved after it originated (*McLean v. Arkansas Board of Education* 1982).

Judge Overton's opinion contains a fairly comprehensive account of the origins of the creationist movement, and of the political machinations involved in getting the Arkansas legislation passed. On the basis of evidence that the legislation was religiously motivated, the judge concluded that it failed to meet the strict standards derived from the establishment clause. In addition, however, Judge Overton held forth on the essential characteristics of science (*McLean v. Arkansas Board of Education* 1982, IV [C]).

(1) It is guided by natural law;

(2) It has to be explanatory by reference to natural law;

(3) It is testable against the empirical world;

(4) Its conclusions are tentative, i.e. are not necessarily the final word; and

(5) It is falsifiable.

He opined that creation science failed to meet these criteria on all counts.

The next move for the conservative Christian anti-Darwinists was to more scrupulously expunge all theological references from their alternative story of the world's history. Thus was born the "intelligent design" movement, which in some sense grew from Phillip E. Johnson's book *Darwin on Trial*, published in 1991. Johnson, a University of California, Berkeley, law professor with no evident scientific background, employed all his lawyerly skills in reiterating arguments against evolutionary thought dating back to the time of *On the Origin of Species*. This was followed by other books, notably Michael Behe's *Darwin's Black Box* (1996), works by William Dembski (1999, 2002), and additional writings by Johnson (2000). Behe, a biochemist, and Dembski, a mathematician, present arguments that life's observable complexity cannot be accounted for by any form of naturalistic evolutionary process, particularly one of random mutations coupled with natural selection. It follows that there must have been design by an intelligent agent at work at some stage or stages in the historical processes leading to the present.

Intelligent design provided a new basis for attacking the teaching of evolution in the public schools. Working mainly through the offices of the Discovery Institute, a conservative think tank, intelligent design advocates have striven to mandate that the argument be taught along with evolution as an alternative, scientifically acceptable theory, free of the taint of religious motivation. This new generation of anti-Darwinists is generally inclined to accord a role for evolution; they just do not want to leave out a role for the intelligent designer. Their opposition is to a naturalistic system that envisions only natural forces working over time on the material substance of the world.

Antievolution legislation is a continuing feature of the agendas of many state legislative bodies, as well as state and local school boards.[3] The Dover, Pennsylvania, school board actions and the resulting legal challenges provide a recent example. Dover's local school district board in December 2004 instructed teachers to inform students of gaps and other problems with the theory of evolution, and in addition to tell them about other theories of evolution, "including, but not limited to, intelligent design." The district's seven biology teachers refused to follow these instructions, so the district's two top administrators went around to biology classrooms and read students a brief statement explaining that Darwinism is only a theory, and pointing to books in the school library that could inform the students about intelligent design, particularly the book *Of Pandas and People,* written by Percival Davis and Dean H. Kenyon (1989). Eleven parents of children in the school district brought suit on the grounds that teaching intelligent design constitutes an unconstitutional establishment of religion. The case was tried before judge John Jones III of the District Court for the Middle District of Pennsylvania.

The Dover suit became a big deal, and was widely reported in national news media (Talbot 2005). The plaintiffs lined up many experts in science education, theology, and philosophy, as well as scientific experts on evolution. Michael Behe, the aforementioned author of *Darwin's Black Box,* and perhaps the leading scientific proponent of intelligent design, testified for the defense. Judge Jones issued a 139-page decision in which he found that intelligent design fails to qualify as science "on three different levels, any one of which is sufficient to preclude a determination that ID [intelligent design] is science. They are: (1) ID violates the centuries-old ground rules of science by invoking and permitting supernatural causation; (2) the argument of irreducible complexity, central to ID, employs the same flawed and illogical contrived dualism that doomed creation science in the 1980's; and (3) ID's negative attacks on evolution have been refuted by the scientific community" (*Kitzmiller et al. v. Dover Area School District, et al.* 2005, 67).

In a section entitled "Whether an Objective Dover Citizen Would Perceive the Defendants' Conduct to Be an Endorsement of Religion," Judge Jones recited abundant evidence that the defendants acted from religious motivations, and that these were clearly evident in the materials provided

3. For information on efforts active in 2005, see http://www.livescience.com/strangenews/ 050927_ID_cases.html.

to parents and others in the district. He was forceful in his assessment that certain members of the school board had lied about the extent of their religious motivations in establishing the policy regarding the teaching of intelligent design, and about their contacts and involvement with the Discovery Institute. Finally, the judge addressed the argument that if there is controversy over the correctness of evolution versus another model, such as intelligent design, then it is appropriate to teach the controversy: "ID's backers have sought to avoid the scientific scrutiny which we have now determined that it cannot withstand by advocating that the controversy, but not ID itself, should be taught in science class. This tactic is at best disingenuous, and, at worst, a canard. The goal of the ID movement is not to encourage critical thought, but to foment a revolution that would supplant evolutionary theory with ID" (*Kitzmiller et al. v. Dover Area School District, et al.* 2005, 89).

These are strong words. It would appear that science has successfully met a challenge to its authority. But just what kind of authority is involved here, and who actually accepts it? It is certainly not accepted by the proponents of intelligent design, who lost no time in publishing a critical analysis of the judge's decision (DeWolf et al. 2006). The scientific community apparently convinced Judge Jones to apply criteria for distinguishing science from pseudo-science that effectively maintained the validity of scientific authority. It must be remembered, however, that the Dover decision is binding only in the Middle District of Pennsylvania. Because it has not been appealed to and upheld at an appellate level, no other federal district judges are bound by it, though it may have persuasive authority with other judges as they formulate their opinions in similar cases. The widely reported decision may influence the views of many in the general population, but we do not know that this is so. Meanwhile, school boards across the country are introducing requirements that in effect challenge the authority of science to decide what should be taught as science in public school classrooms. Certainly not all the mandates to teach that evolution is "only a theory, not proven," that it is incomplete or inconsistent in many respects, or that it must be taught along with intelligent design or some other theologically linked version of evolutionary development, will be challenged in the courts. The conflicts that arose in Dover, Pennsylvania, are being repeated across the country because conservative Christians have gained control of many school boards that control curricular changes.

In November 2005, only four days after closing arguments, and about a month in advance of Judge Jones's decision, the voters of Dover voted in a new slate of school board members. All eight school board members running for reelection were ousted in favor of a slate of candidates running under the Dover Cares banner. The campaign was hard fought and divisive. In a mass mailing, the incumbents alleged that the Dover Cares slate was allied with the American Civil Liberties Union, which they associated with various unpopular causes. The Dover Cares campaign, on the other hand, argued that the incumbents had brought down unwelcome publicity and disrepute on the community. Although the Dover Cares slate was swept into office, the election was close. One of the seats was decided by only 26 votes out of 5,058 votes cast. Judging from these results, it seems fair to say that the authority of science in opposition to that of conservative religious beliefs is not strongly upheld by the citizens of Dover.

Meanwhile, in April 2005, Jerry Johnson, the conservative pastor of the First Family Church in Overland Park, Kansas, delivered a sermon attacking evolution, putting it into the same basket with abortion and gambling. He is reported to have said, "Getting intelligent design into school curricula is the worthiest cause of our time and the key to reversing the country's moral decline. The evangelical and ID communities must work together to make that happen" (Bhattacharjee 2005a). Kansas has for years been the scene of a seesaw battle over the science curriculum. The arena in which much of the action has occurred is the Kansas State Board of Education, which has the power to set the science standards for all state school curricula. Johnson has a good chance of having his way. In the November 2004 election, evangelicals and intelligent design proponents, led by John Calvert, a manager of the Intelligent Design Network centered in Shawnee Mission, Kansas, were successful in establishing a conservative majority on the state board. As the board prepared its new draft standards, it was clear that they would be modified to redefine what constitutes good science so as to both accommodate intelligent design and cast doubt on mainstream evolutionary theory. In anticipation of the outcome, the National Academy of Sciences and the National Science Teachers Association denied the board permission to employ language from their publications dealing with education standards, on the grounds that the standards proposed by the Kansas board inappropriately single out evolution as a controversial theory, and that its definition of science blurs the distinction between scientific and other ways of understanding (Bhattacharjee 2005b). The AAAS weighed in by supporting the groups' copyright move against the Kansas board. The denial of copyright required the Kansas board to rewrite a good deal of material that is conventionally lifted from the standards promulgated by the two organizations. In the August 2006 primary elections, moderate candidates for the board regained sufficient ballot spots to ensure that they would have a slim majority following the fall elections. The closeness of the elections and the seesaw pattern of majority control of the school board reflect a split in Kansas society regarding conservative versus moderate stances on a variety of topics, but the teaching of evolution and related science topics (such as the age of the earth) are dominant concerns. The state board of education elections were not prominently in the public eye in 2008. Perhaps because they sensed that Kansas voters are weary of seeing the board being ridiculed in the national media for some of its prior actions, none of the elected candidates voiced an intent during the campaign to overturn the board's present moderate stance. Nevertheless, there is continual pressure for standards that place evolutionary theory in doubt and include intelligent design in the curriculum (Bauer 2008).

The Kansas example is yet another piece of evidence that, in the matter of public school education, the authority of science compared to religion is weak. At its 2006 annual meeting in St. Louis, the AAAS held a one-day event, "Evolution on the Front Line," for K–12 science teachers, which was also open to students, policy makers, scientists, and others. The event was designed to give teachers a voice on the evolution issue, and to indicate how they can best be supported. One point made by many of the teachers is that evolution is frequently not taught because it is not assessed by statewide standardized testing. In fact, only a few states across the country routinely test knowledge of human evolution. Thus, statewide testing programs form an indirect avenue for suppressing the teaching of evolution, one of the bedrocks of the biological sciences.

Why does science have such a tough time of it with respect to religion in the matter of teaching evolution in the public schools? Quite simply, it is because most people entertain beliefs that are inconsistent with mainstream science. In 2004, CBS News conducted a telephone poll among a nationwide random sample of 885 adults, among whom 795 were registered voters. Below are the percentages of "yes" responses to the questions listed (CBS News 2004):

| | |
|---|---|
| God created humans in their present form | 55% |
| Humans evolved, God guided the process | 27% |
| Humans evolved, God did not guide the process | 13% |
| Favor schools teaching creationism and evolution | 65% |
| Favor schools teaching creationism instead of evolution | 37% |

The survey was repeated in 2005, with more or less the same results. Interestingly, the results for the first question are pretty much the same when a specific time period was included, such as, "God created humans in their present form within the last ten thousand years." These results suggest that ignorance or outright rejection of the findings of mainstream science is widespread. Further, the polls showed that white evangelical Christians are more disposed to believe that an accommodation between their religious beliefs and science is not possible. The powerful influences of biblical literalism and political activism on beliefs about evolution and human origins are revealed in a study published by the National Center for Science Education. The public acceptance of evolution is lower in the United States than in any of the thirty-four nations surveyed except for Turkey, and substantially lower even than in traditionally Catholic countries such as Portugal or Ireland (Miller, Scott, and Okamoto 2006). Statistical analyses showed that the effect of fundamentalist religious beliefs on attitudes toward evolution was nearly twice as great in the United States as in a group of nine European countries.

One might question whether the polls are perhaps tapping into a less-educated segment of the population. Although better-educated people are more inclined to endorse evolution, nonscientific beliefs are commonplace among college students. For example, professor Will Provine, a geneticist, for several years has been surveying the beliefs of his Cornell University undergraduate students in an evolution course for nonbiology majors (Holden 2006a). He says that about 70 percent of the students hold creationist beliefs along a spectrum that ranges from biblical literalism to a conviction that human existence could not have come about without divine intervention. Furthermore, surveys of students during and at the end of biology courses dealing with evolution show that students do not readily change their beliefs.

All of these results together are evidence that science lacks cultural authority with respect to religious beliefs. Simply laying out the evidential basis for belief in a Darwinian story of how the world has come to be as we find it will not change deeply held convictions. It is not just a matter of whether to accept the Genesis story of the Bible; people fear that they will lose the religious faith that informs their daily lives. Science takes a backseat to religion in the intellectual formations of many people. Some find themselves able to move away from naive religious beliefs that are inconsistent with well-established scientific evidence and theory. Far more, however, never come to a point where such a contest of ideas even occurs.

Surveys reveal a low level of genetic literacy in the population at large. Fewer than half American adults can provide a minimal description of DNA. Even among those who do possess some genetic literacy, however, the idea of human exceptionalism is deep-seated (Miller, Scott, and Okamoto 2006). In most peoples' thinking about such matters, the factors discussed in chapter 1 that provide a justification for the authority of science simply do not operate. Without a vast change in the content of public school curricula, we are consigned to continuing to live with more of the past century's skirmishing between science and religion. In this continuing struggle, science lacks the opportunities and power available to religion to inculcate children with its core ideas and ethos.[4]

Evolution is one of the core ideas in a modern scientific conception of the world, but it generally does not play a role in important decisions people need to make about their daily lives. How do conflicts between science and religion play out when the stakes are social policies that do importantly affect such decisions?

## THE SCIENCE AND THEOLOGY OF HUMAN REPRODUCTION

The Vatican—or, more formally, the State of the Vatican City—is the smallest independent nation in the world. Created by the Lateran Treaty of 1929 between the Kingdom of Italy and the Roman Catholic Church, it consists of about 108 acres (44 hectares) within the city of Rome. Among its many terms, the treaty stipulates that Roman Catholicism is the religion of Italy, and makes a financial settlement with the church in compensation for the loss of its sovereignty and dominion over land dating from formation of the Kingdom of Italy in 1870. (A 1984 amendment to the treaty ended the church's status as the state-supported religion of Italy.) The treaty brought to an end a long period in which the papacy existed in an uneasy and uncertain relationship to the Italian government.

The loss of the so-called Papal States in the period around 1860 represented the final stage in the Catholic Church's loss of temporal power. For Pope Pius IX, who had ascended to the papacy in 1846, it was a bitter

---

4. In 2006 the AAAS published *The Evolution Dialogues: Science, Christianity, and the Quest for Understanding* in an attempt to address misunderstandings of evolutionary science and its relationship to Christian beliefs (AAAS 2008a). See also *Science, Evolution, and Creationism* (National Academy of Sciences 2007b). These publications are addressed to nonscientists, and particularly to science teachers at all levels.

outcome, the culmination of a series of setbacks in his prolonged struggle to maintain his hegemony over political territories (Kertzer 2006). When Victor Emmanuel seized Rome and made it the capital of a united Italy in 1870, he proposed and enacted into law a plan that accorded the pope the rights of a sovereign, an annual remuneration, and the rights to the Vatican palaces, the Lateran palaces, and the summer residences at Castel Gandolfo. Pius IX refused the offer, and instead chose to view himself as a prisoner within the Vatican. Nor were his troubles at home the end of his travails. Various concordats—agreements between the pope and nations for the conduct of church affairs—were abrogated, and other infringements on church authority occurred in many nations, both Catholic and Protestant.

Following all of these political setbacks, and in the face of rapidly changing social conditions, including political, philosophical, and cultural movements seen as threats to traditional religious beliefs, Pius IX settled into a very conservative outlook. In response to the loss of most of his political authority, the pope moved to more visibly exercise his religious authority. Among his ecclesiastical actions was his 1854 proclamation of the Immaculate Conception of the Blessed Virgin as a dogma of the church, and his convocation of the Vatican Council in 1869. During that Council, papal infallibility was also made a dogma of the church. The Immaculate Conception dogma pertains to Mary, the mother of Jesus. It entails that at the moment of her conception, the soul of Mary was formed free from the taint of original sin, whereas all other humans conceived since the fall of Adam and Eve are deemed by the church to be conceived with that stain, and thus liable to a host of human frailties. The dogma of papal infallibility entails that when speaking ex cathedra (in the exercise of his office) with respect to a doctrine concerning faith and morals to be held by the entire church, the pope speaks infallibly, "and that such definitions of the Roman pontiff are of themselves, and not in consequence of the Church's consent, irreformable" (All Catholic Church).

The Beginnings of Human Life

The Immaculate Conception dogma is relevant to our story because it reveals something of the Church's position on the question of when human life begins. In Roman Catholic doctrine human life begins with the creation of a soul, an immaterial entity defined as the ultimate internal principle by which we think, feel, and will, and by which our bodies are animated. The soul is created at the moment of conception; thus, a human life has begun when fertilization has occurred (Shannon and Wolter 1990). This idea, referred to as immediate animation, is not precise in the light of modern biology, which reveals the process of conception to be complex and extended in time. Nevertheless, it places the time of animation, the moment of "ensoulment," close to the time of sexual intercourse. In earlier times the Catholic Church held varying views about the beginnings of human life. The theologians of the Council of Trent (1545–63) asserted that no human embryo could be infused with a human soul except after a certain period of time. Later theologians argued that ensoulment must occur at the time of conception (U.S. Conference of Catholic Bishops n.d.). Pius IX's enshrinement of the Immaculate Conception in dogma served to reinforce the belief in immediate animation. It has remained church doctrine that the beginning of human life coincides with the act of conception. The church is, of course, implacably opposed to abortion at any stage in the development of a fetus. It is also opposed to any steps that might be taken to frustrate conception—that is, to contraception in any of its various forms.

It is arguable whether any increased knowledge of human biology could influence the Catholic Church's position on the beginnings of human life, and hence on questions relating to contraception or abortion. Nevertheless, it is germane to our discussion that scientific research has revealed many aspects of the process of conception that could inform thoughtful consideration of the moral aspects of contraception, abortion, use of stem cells in research, and much more. We now know that it does not make scientific sense to talk of a "moment of fertilization." When male sperm are released into the female reproductive tract, they must travel up the fallopian tube to join with the egg, a process that takes about ten hours. The interaction of the sperm with the egg is a complex biochemical process that culminates after some time in the sperm reaching the inner portion of the egg. Another twenty-four hours or so are required for the chromosomes (the entities containing the genetic material) of the egg and sperm to meet to form an organism called the zygote. But it cannot be said that at this point a unique individual has been formed, because the zygote is capable of splitting into two or more zygotes at any point over the following fourteen days or so. In addition to these considerations, quite a large percentage of fertilized eggs fail to continue to develop, but simply abort spontaneously. Zygotes that survive past the first twelve days or so are termed embryos, though it must be said that the attachment of names to entities in this field is fraught with controversy. The embryo then begins the process of gastrulation, in which the previously undifferentiated cells of the zygote begin to move and take on differentiated roles. For some, this stage represents the real beginning of a unique human life. While the human embryo has the beginnings of basic neural circuitry at about week eight, neurological development in the cerebral cortex does not occur in at least the first twelve weeks. Significant development of neural pathways critical for rational thought only occurs in the final trimester of gestation (Larsen et al. 1993; Morowitz and Trefil 1992).

The attitudes of individuals with respect to such highly charged topics as abortion and contraception derive from their moral and ethical outlooks and religious beliefs. For some, what science has to say about the biology of human reproduction is irrelevant. Those who fully subscribe to the tenets of Roman Catholicism or positions advocated by Christian political action groups such as the Moral Majority (founded in 1979, dissolved in 1989), the Christian Coalition, or any one of several other conservative religious denominations are persuaded through their submission to religious authority of what is moral and right with respect to issues of human reproduction. For others, scientific knowledge regarding cell biology and human reproduction may contribute to the formation of consciously constructed opinions on the various issues that arise. The authority relationships between science and religion in this arena are therefore complex. In an important sense, the expert authority of science, which tells us how things appear to be in the biological world, exists apart from the traditional, legal-rational, and moral authority of religion, which largely dispenses with the facts of biology in propounding a doctrine and commanding behavior in consonance with it. Nevertheless, what science has to say may affect the willingness of some to accept religious authority if they conclude that some scientific findings are relevant and serve to weaken religious authority. We can best explore this notion by examining in some detail how the Catholic Church's positions on contraception have played out in modern American culture.

## The Church on Contraception

The Catholic Church has consistently maintained its opposition to contraception. So long as contraception was not all that easy to effect, it was not a topic that received much attention. Things began to change with the advance of science, particularly the fields of medicine and chemistry. One of the most extensively used contraceptive devices is the condom, a penile shield. An early form of the condom was made from linen, another from animal intestine. They were seen to have value in preventing the spread of venereal diseases. In the sixteenth century an epidemic of syphilis was spreading into Europe, and Naples, Italy, was its epicenter. Gabrielle Fallopius, an Italian physician, claimed to have conducted a trial of the condom as a prophylactic device; he had 1,100 men use a sheath made of linen. Though condoms might be effective, their use was not widespread, because they were expensive and difficult to use. The invention of vulcanized rubber in the latter part of the nineteenth century, however, led to the latex condom. By 1930 it was possible to produce relatively inexpensive condoms that were almost as thin as those available today. Their ready availability posed a challenge to church authority; their use was forbidden because they interfered with the possibility of conception following sexual intercourse.

Adding to papal concerns, the traditional family seemed to be in danger due to a variety of changes in societal norms. Although divorce was uncommon in the nineteenth century in Europe and the United States, divorce rates rose steadily in the twentieth century (Furstenberg 1994; Wojdacz n.d.). During the 1920s an air of postwar decadence pervaded much of Western society. In the United States the Jazz Age culture seemed to say that anything goes. In Europe the Depression hit hard, particularly in Germany, reeling under the burden of reparations for World War I. *The Threepenny Opera*, written by dramatist Bertolt Brecht and composer Kurt Weill, was the biggest hit of the 1920s in Berlin; it gave voice to the cynicism and despair of the times, and questioned the conventional morality espoused by the church in the face of the hunger and deprivation of the working class. Faced with these disturbing trends, Pope Pius XI issued a papal encyclical—a letter to cardinals, archbishops, bishops, and other select church leaders—on the subject of Christian marriage, which appeared at the end of 1930. In his long letter, *Casti Connubii*, the pope inveighed against a long list of sins against the sacrament of marriage, including abortion, divorce, and conception: "But no reason, however grave, may be put forward by which anything intrinsically against nature may become conformable to nature and morally good. Since, therefore, the conjugal act is destined primarily by nature for the begetting of children, those who in exercising it deliberately frustrate its natural power and purpose sin against nature and commit a deed which is shameful and intrinsically vicious" (Pius XI 1930, sec. 54).

*Casti Connubii*, which is generally conceded to be the work of the prominent Jesuit moralist Arthur Vermeersch, probably carried some sense of urgency because, at the same time, the Church of England at its triennial Lambeth Conference in 1930 gave its official, though somewhat guarded, endorsement of contraception. The encyclical stood for more than thirty years as the Catholic Church's official stand on issues of marriage and human reproduction. Some conservative theologians were of the view that it constituted an infallible pronouncement. Others saw it as simply a re-capitulation of traditional teaching, possibly subject to reconsideration. Not long after it had been issued, science presented the church with yet another challenge to traditional understanding of human reproduction. Two medical researchers, Kyusake Ogino in Japan and Hermann Knaus in Austria, established the existence of a "safe period" around the five or six days during the menstrual cycle when ovulation occurs. By avoiding sex-ual intercourse during those days a couple could, at least in principle, avoid conception. Vigorous discussions among theologians over whether the practice of the "rhythm" method was acceptable culminated in 1951 when Pope Pius XII gave his qualified approval for its use (Noonan 1965, 518). But yet another challenge to ecclesiastical authority soon appeared, in the form of the birth control pill. Gregory Pincus, an authority on mammalian reproduction, and M. C. Chang, a senior colleague at the Worcester Foun-dation for Experimental Biology, jointly developed an oral contraceptive that prevents ovulation by mimicking the action of the female hormone progesterone. They teamed up with Dr. John Rock, who operated a repro-ductive clinic in Massachusetts. Human trials in Puerto Rico and Haiti demonstrated the successful use of "the pill." In 1960 the Food and Drug Administration authorized marketing of the drug for miscarriage and cer-tain menstrual disorders.

The pill created new and vexing issues for the theologians to sort out. When administered in the right way it could improve the chances of con-ception, by regulating the menstrual cycle. But conversely it could also act as a contraceptive. Although many Catholics practiced the rhythm method to limit family size, many found that irregularities in the menstrual cycle led to a significant prospect of unwanted pregnancy. Rhythm practice was referred to in some circles as "Vatican roulette." During this time there was a growing sentiment among many Catholic theologians to move away from the strict and narrow view of sexual union in marriage as justified solely by God's command to go forth and procreate. Instead, the sex act could be understood in a larger sense, as having a depth and gravity arising from shared love, a means of binding the two marital partners together. Procreation could be seen then as flowing in a natural way from marital love. Thus there was a feeling of change in the air when Pope John XXIII convened the Second Vatican Council in 1962.

Among the reformist clergy, there was a fear that the implacable opposition to any form of conception embodied in *Casti Connubii* would find its way into Vatican II doctrine. Liberal elements in the church convinced the pontiff to appoint a special commission, the Pontifical Commission for the Study of Population, Family, and Births. By the time the commission held its first meeting in 1963, however, Pope John had been dead for four months. The Commission was continued under the papacy of Pope Paul VI. Each successive meeting involved a greater number of participants, taken from increasingly broad segments of the Catholic community. The fourth and final meeting, held in 1966, involved lay people, including a few married couples. There was a strong feeling in this group that important changes needed to be made. Robert McClory describes a key episode, in which the theologians in the group expressed their views: "De Riedmatten [a Vatican diplomat moderating the session] asked each of the nineteen theologians to give a six-minute presentation of his position, after which a vote would be taken on two questions: Is the doctrine of *Casti Connubii* irreformable? And is artificial contraception an intrinsically evil violation of the natural law? The result was no on both questions—both by 15–4 tallies. Though de Riedmatten insisted the vote was strictly provisional and not to be considered absolutely final, everyone realized its significance" (McClory 1995, 99).

In the debates that followed, the reformists alluded to new understandings of human biology, of the institution of marriage, and of the place of humans in the world—in short, of largely scientific advances—that cast the questions facing the church into terms different from those that seemed appropriate in the past. The conservatives, by contrast, attacked these revisionist thoughts as damaging to the church's authority and claims of infallibility. The papal Commission finished its work in June 1966, and issued what came to be called the Majority Report, delivered later that month to Pope Paul. But the Vatican conservatives were not yet vanquished. Under the leadership of the elderly Vatican insider Alfredo Cardinal Ottaviani, a small group of dissidents prepared a "Minority Report." They convinced the pope that he should reconsider his initial favorable reaction to the Majority Report. Pope Paul was faced with deeply conflicting theological arguments and opposing outlooks on the church's role in the world. During the time that he agonized over the decision that faced him, he held an interview with French philosopher and longtime friend Jean Guitton.

Citations of the pope's words at that meeting indicate just how edgy he had grown:

> One should not say that the Church is in a state of doubt or is not sure of itself. The Church is well informed on all the latest data about contraception. Maybe something has changed? In that case, let's see what science has discovered that our ancestors didn't know. Let's see if something should change in the law. Put questions to the scientists. But having listened to them, let's listen to the voice of conscience and the law. These must make demands, raise the level. Any attenuation of the law would have the effect of calling morality into question and showing the fallibility of the Church which then, like the Jewish synagogue, would have imposed too heavy a yoke on people. . . . Theology would then become the servant of science . . . science's handmaid, subject to change with each new scientific discovery so that tomorrow, for example, we would have to admit procreation without a father; the whole moral edifice would collapse . . . Who is to say that another scientific discovery will not come along that will subvert the discovery of the pill, showing, for example, that the pill may produce monsters in the next generation? (McClory 1995, 131–32)

A long period of waiting for the pope's final stand came to an end on July 29, 1968, with the issuance of a new papal encyclical, *Humanae Vitae*. The conservatives had won; the document reaffirmed the prohibition on contraception spelled out in *Casti Connubii*. The new encyclical dumfounded all those who were hoping for and expecting a more nuanced, open, and permissive approach to contraception. Liberal theologians in both Europe and the United States were outraged, particularly by the encyclical's seeming reliance on an Augustinian view of love as justified only by procreation, to the exclusion of a more personalist image, a mutual giving of one's individuality.

This tale has many interesting themes, not the least of which is the church's obsession with maintaining its authority. Although papal encyclicals do not formally carry the stamp of infallibility, many of them take on that status de facto because of their import. The conservatives in their discussions with Pope Paul argued that to make any concessions on contraception that could be seen as contradicting *Casti Connubii* and other, earlier papal statements, would, in effect, trash the doctrine of papal infallibility and undermine papal authority. Ironically, the obsession with authority, coupled with the traditionalists' idiosyncratic readings of the earlier church fathers such as Augustine and Aquinas, has had the effect of vastly diminishing the authority of the church in matters related to human reproduction (G. Wills 2000).

Abundant survey data show that since the publication of *Humanae Vitae*, Catholics worldwide have increasingly practiced contraception in violation of church edicts (Althaus 1991; Fehring and Schmidt 2001; Catholics for Choice 2008). For example, among women between the age of fifteen and forty-nine in the United States, 70 percent of Catholic women were using some form of contraception in 1995, about the same percentage as all women in that age group. Women who had attended Catholic school exhibited patterns of contraceptive use that were about the same as those without religious education. Strikingly, sterilization (most of the time by women) was by far the most frequent method of contraception employed by Catholic couples, with an oral contraceptive ranking second and condoms third. All three of these methods are, of course, explicitly forbidden by the church. In European countries, many of which are much more heavily Catholic than the United States, the birth rates are below replacement levels. The case of Ireland is particularly interesting. In 1979 legal restrictions on contraception were lifted. The birth rate in Ireland was about four children per couple in 1970; by 1995 the number was below two (Bloom and Canning 2004). Similar trends are also evident in Latin America. The data are quite clear: the Catholic Church has very limited moral authority with Catholic laity in matters of sexual behavior and reproductive health.

A second interesting theme is embedded in Pope Paul's interview with Jean Guitton. It reveals his suspicions of science as a source of good, and a fear that the church will not be able to deal effectively with the complex moral and religious issues emerging from scientific advances. The pope expresses the fears of the Vatican hierarchy—a collection of celibate, rather aged men not in good contact with the realities of modern life—that any authority the church cedes to science is authority lost.

It is important to see that at the biological level, the issue of contraception is not one in which the authority of the Catholic Church and science are in direct opposition. There is no real question of the expert authority of science in matters related to contraception, even though pseudo-science may be invoked for propaganda purposes (for example, by antiabortionists). At the level of the biology associated with human fertility, science's role has been that of the expert. The church asserts that a human soul is formed at the time of conception. The church is free to define the time of conception in any way it likes, because the use to which it puts that definition does not fall within the domain of science. Scientific findings make it clear that "time of conception" has no clear biological meaning. Conception and fetal development are biological processes, not ecclesiastical definitions. A teaching that is uninformed by any of what science knows about these most vital biological matters is bound to suffer some loss of moral authority with Catholic laity. Although, at the most basic level, science and the Catholic Church are not at opposite poles, neither do they reside comfortably alongside each other.

Although the Catholic Church has little moral authority with its lay members in contraceptive matters, it has labored mightily to exert institutional influence in society more generally, and this has led to conflicts between the church's traditional authority and the authority of science. In the United States, the Conference of Catholic Bishops serves as a political arm of the Vatican. In 1975 the American Catholic bishops, at their annual meeting, laid out a Pastoral Plan for Pro-life Activities. This political document called for influencing political processes at local, state, and national levels. In addition to its attempts to exercise moral authority with Catholic laity, the church also participated—indeed, led—in the creation of such Christian right organizations as the Moral Majority and, later, the Christian Coalition. Catholics were central in the formation, support, and leadership of these organizations.

In 1976, following the 1973 *Roe v. Wade* decision of the Supreme Court that affirmed a woman's right to choose abortion under certain circumstances, Rep. Henry Hyde (R-Ill.) inserted into the Medicaid portion of the funding legislation for Social Security an amendment providing that "none of the funds provided by this joint resolution shall be used to perform abortions except where the life of the mother would be endangered if the fetus were carried to term; or except for such medical procedures necessary for the victims of rape or incest when such rape or incest has been reported promptly to a law enforcement agency or public health service."[5] Henry Hyde grew up in an Irish Catholic family in Chicago, and maintained strong ties with the church throughout his long political career. He consistently maintained a strongly antiabortion stance in his congressional activities.

5. This is the language of the amendment to the 1980 legislation; it varied in detail from year to year following the initial amendment (Hyde Amendment, Pub. L. No. 96–123, 109, 93 Stat. 926).

The Hyde Amendment constitutes a significant retreat from the spirit of *Roe v. Wade* by effectively banning the use of federal funds for abortions. It was challenged in the courts by pro-choice advocates. Lower courts ruled against the government, but eventually, in 1980, the Supreme Court reversed, ruling in *Harris v. McRae* that the Hyde Amendment was constitutional as it stood. A brief of amicus curiae urging reversal of the earlier rulings was filed in *Harris v. McRae* on behalf of the United States Conference of Catholic Bishops. The Conference maintains an office of legal counsel, which regularly files amicus briefs in cases involving contraception and abortion.

One of the Conference's publications is entitled *Ethical and Religious Directions for Catholic Health Care Services* (U.S. Conference of Catholic Bishops). The guidelines, which apply to the nearly six hundred Catholic hospitals, as well as Catholic health care facilities, in the United States, mandate against (a) prescribing or dispensing birth control devices or medications; (b) counseling on contraception practices or the use of condoms to prevent the spread of HIV/AIDS; or (c) any sterilization, abortion, or in vitro fertilization procedures. For some, a Catholic hospital may be the only accessible health care facility, regardless of whether they are Catholic. It is not so clear, however, that the Conference guidelines are scrupulously followed in all locations. The hospitals are generally administered by religious orders of nuns or priests, and the chain of responsibility often does not pass directly through the local bishop or archbishop. As just one example of a departure from the guidelines, a 2005 survey of staff at Catholic hospitals found that about 50 percent of them will, upon request, dispense the abortifacient "morning after pill" to rape victims, in contravention of the guidelines laid down by the bishops (Catholics for Choice 2006).

The foregoing examples illustrate the church's continuing attempts to exercise its authority through institutional mechanisms, including politics. These efforts have given rise to conflicts between science and religion in which science has a direct stake in the issue at hand—most notably, in connection with stem cell research.

## STEM CELLS

Stem cell research has given rise to the most complex and vexing challenges to the authority of science seen in our time. The challenges are not simply in the obstacles to achieving effective medical therapies based on stem cells, though there are plenty of those. Stem cell research has given rise to many moral questions relating to the ethics of pursuing particular research questions, including issues of intellectual property rights, scientific competition and fraud, governmental control of science, and conflicts between scientific goals and religious beliefs. Some of these matters will be dealt with in other sections of this book. In this chapter I want to focus primarily on conflicts between science and religious authority. First, a short primer on stem cell science. I should say in advance that by the time you are reading this book, it is likely that much will have changed in the scientific understanding I describe. If in fact that is the case, however, it will not alter the pertinence of my story, which is to show how at a particular point in time scientific knowledge and religious authority have intersected.

In chapter 1 we considered assisted reproductive technologies (namely, in vitro fertilization) to illustrate a simple distinction between expert authority and moral authority. To recapitulate, cells taken from the female are fertilized by sperm cells taken from the male in a procedure occurring in a petri dish—hence the name in vitro ("in glass") fertilization (IVF). After two to three days, one or more of the fertilized eggs that has developed into an embryo, the first stage in formation of the fetus, is implanted in the woman's uterus. Or, in a variation on this procedure, the embryo is allowed to further develop in the petri dish for an additional day or two, forming what is called a blastocyst, a more organized but still very small collection of cells. What was not said explicitly in chapter 1 is that either procedure generally leaves a number of female eggs, embryos, or blastocysts unused. Typically these are frozen under conditions that would permit their use at a later time, if needed. Normally the need for the unused eggs, embryos, or blastocysts does not arise; thus, they have accumulated in large numbers in fertility clinics. It is estimated that there are at least four hundred thousand embryos or blastocysts in fertility clinics around the country, although only eleven thousand of them have been designated for research purposes by their donors.

The relatively few cells that make up the embryo or the interior of the blastocyst at this stage of development are undifferentiated. That is, they have not as yet taken on the character of a particular kind of cell, such as a skin cell, liver cell, and so on, but they have the capacity to do so under the right circumstances. Such cells from blastocysts are termed *pluripotent* to indicate their capacity to differentiate into any one of a number of different cell types. The process by which the cell differentiates to become a specific kind of cell occurs at a later stage in development. In recent years it has been claimed, and widely believed, that pluripotent (or embryonic) stem cells have great biomedical potential. In 1998 James Thomson at the University of Wisconsin successfully removed embryonic stem cells from human blastocysts and cultured them in the laboratory. Around the same time, John Gearhart and colleagues at Johns Hopkins University carried out similar work on stem cells derived from fetal tissue. Both groups demonstrated that their cells could proliferate in an in vitro culture and maintain their capacity to differentiate into various types of tissue cells. The potential for therapeutic biomedical applications is just this: If such cultured embryonic stem cells could be induced under the correct conditions in the laboratory to form cells of a particular kind, say pancreas cells, it might be possible to employ stem cell cultures to replace cells that have died in a patient. They might, therefore, have the potential to cure diseases such as diabetes, Parkinson's, and others that result from the death or diminished vitality of particular organs or kinds of cells.

The biomedical science in this area, termed "regenerative medicine," is at a very early stage. To date no cure for any major disease has been found by this route. It is fair to say that much basic research needs to be done if science is to understand and control the many processes that occur as cells undergo the transition from stem cell to differentiated cell. Among the scientific issues requiring further study is just how long stem cell cultures can continue to proliferate. There is evidence that genetic alterations creep into the cultures over time (Maitra et al. 2005; Johns Hopkins Medicine 2005). As they multiply many times, they accrue mutations. Human stem cell cultures also tend to form teratomas (benign cancerous growths).

Stem cells are also found in the organs of a mature adult, though they are typically not numerous, nor easy to identify. For example, in bone marrow there are cells that have the capacity to give rise to red or white blood cells. These stem cells, however, are not normally capable of forming cells characteristic of other bodily organs, such as the liver. They have already undergone programming, as it were, so that they are able to form only blood cells. In limited cases it has been possible to coax adult stem cells to reverse some of their programming so that they could be used to grow cells characteristic of other organs. If stem cells could be extracted from a person who suffers from a disease, be deprogrammed, and then be made to form cells characteristic of the diseased organ, they might be implanted in the person, with the goal of restoring the organ's viability. A great advantage of adult stem cells would be that the body's immune system would not reject the cells. There have been some successes and promising leads (Vogel and Holden 2007; Holden and Vogel 2008), although scientists are divided in their judgments of how effective adult stem cell use can be. It is not easy to coax adult stem cells to undo their programming, and it is not clear just how versatile the resulting cell cultures can be. Furthermore, adult stem cells do not grow in cultures as readily as embryonic stem cells. Exploring the use of adult stem cells for regenerative medicine is very much worthwhile, though it is not clear that it will eventually afford solutions for many diseases.

Still a third direction for research involves somatic cell nuclear transfer (SCNT), the process that was used in forming the famous cloned sheep Dolly. In this process, a nucleus was extracted from a cell taken from the udder of an adult sheep, and implanted in a female sheep's egg cell from which the nucleus has been removed. In other words, the nucleus of the udder cell replaces the original nucleus of the egg. The process causes the new cell to behave very much like a primitive, undifferentiated stem cell. It begins dividing and forms a blastocyst; when the blastocyst is implanted in a uterus, it is potentially capable of developing as a fetus to full term. The overall process, as it was used to form the embryo that became Dolly, is termed "reproductive cloning." The aim is to make a copy of the creature from which the nucleus inserted into the egg cell is derived. Reproductive cloning of animals has been the subject of much research, and some successes have been achieved. At the present stage of development of the science, however, it does not typically lead to healthy clones.

If such a method were applied to human cells, the object would not be reproductive cloning. Indeed, cloning a human being by such means is today far outside science's capacity. Almost no one thinks it would be morally or ethically permissible to attempt it. Nevertheless, the thought of hordes of human clones conjures up nightmarish visions of a future world gone awry. It clouds rational discussion of the procedure, often referred to as "research cloning," in which the process of cell division is stopped at the blastocyst stage in order to harvest a supply of stem cells that could be cultured in vitro and used to form cells characteristic of particular organs. Such cell cultures could be valuable in research on cures for diseases such as Parkinson's or diabetes. Research cloning shares with use of adult stem cells the advantage of avoiding cell rejection in therapeutic applications, thus averting the need for immunosuppressive drugs.

Done one cell at a time, SCNT is laboriously difficult work, however, and is prone to failure. While the method has great potential as a tool in basic biomedical research, its applicability to regenerative medicine seems limited. The particular cellular processes that must occur for it to work at all are not well understood. To date efforts to produce embryonic stem cells from cloned human embryos have not succeeded. Widely heralded claims in 2004 and 2005 by a team of South Korean and U.S. scientists to have carried out successful cloning were found to be fraudulent (Wohn and Normile 2006). The extent and depth of the ethical and fiduciary breaches in this case are truly staggering. In spite of this setback, efforts to clone human cell lines continue in many laboratories around the world.

While stem cell science is changing very rapidly, the three major classes of experimental approaches described above are likely to remain, though the detailed procedures and projected research strategies within each will evolve. In any case, this discussion is about how the appearance of a new arena for scientific study generates societal responses that affect both the expert and moral authorities of science, as well as scientific autonomy. In following the manifold ways in which ethics, religion, economics, politics, and social science have influenced the course of events, we will find that we cannot cleanly separate them into nicely demarcated compartments. It is not easy to tell where religion leaves off and ethics begins, or how to weigh the value of free market forces against social policies related to health care. For the present, I hope to confine the range of discussion to the impact of religious authority, and the ways in which the science establishment has dealt with it.

## THE RELIGIOUS VOICE IN STEM CELL POLICIES

In 1995 the United States Congress passed an appropriation bill for the Department of Health and Human Services (HHS) that came with an attached rider prohibiting HHS from using appropriated funds to support research involving the creation or destruction of human embryos for research purposes. The amendment was inserted at the instigation of Rep. Jay Dickey (R-Ark.), and the legislation was signed into law by president Bill Clinton. It has appeared in the HHS, Labor, and Education appropriation bills in succeeding years in more or less the same form. In his four terms as a United States congressman, Jay Dickey's voting record aligned perfectly with the agenda of the National Right to Life Committee.[6]

6. See Congressional Report Cards, http://www.vis.org/crc/getGroups.aspx.

Following the 1998 research that demonstrated the capacity to grow human stem cells in cultures, the question arose as to whether federal research funds could be employed in stem cell research. The Dickey Amendment has meant that HHS could not itself support research that derives stem cells from embryos. It was determined, however, that HHS could support research using stem cells derived through private funds. In 2001, president George W. Bush announced that, for the first time, federal funds could be used to support research on human embryonic stem cells, but that it would be limited to "existing stem cell lines, where the life and death decision has already been made" (White House 2001). Because the cell lines actually available for useful research turned out to be few in number, research in many U.S. laboratories has proceeded with private funding. Thus, for example, in 2004 Douglas Melton and colleagues at Harvard University announced that they had isolated seventeen new human embryonic stem cell lines based on research funded by private dollars (Howard Hughes Medical Institute 2004; Cowan et al. 2004). These cell lines were made available to other researchers, but any research conducted with them would have to be kept strictly separate from any federally funded work, a considerable burden in an active research laboratory with multiple funding sources.

The Catholic Church and many other conservative Christian religious groups hold that a human being is created at the moment of conception, and is from that time forward entitled to all the rights of any other person in society, which of course includes the right to life. Religious conservatives condemn IVF on the grounds that it contravenes God's will that the conjugal act be the sole means of procreation. They also condemn all actions taken with respect to the embryos formed via IVF procedures. The Catholic Church takes these matters very seriously. When Kelly Romenesko, a married teacher at two Catholic schools in Wisconsin and a practicing Catholic, informed her principal in 2006 that the twins she had recently borne had been conceived via IVF, she was fired, and her dismissal was upheld on appeal (ABC News 2006). In 2005, Italian bishops campaigned actively against a referendum to liberalize Italy's very restrictive laws relating to assisted conception, by urging Italians to boycott the polls. The measure fell far short of the required level of balloting, constituting a major victory for the Vatican.

The church is also opposed to human cloning, including research cloning, whether from embryos or blastocysts left over from IVF procedures, or derived from SCNT. Some theologians argue that because the product of an

SCNT procedure in a cell that has all the potential of a cell created in the process of conception, it is ipso facto entitled to the status of a human fetus. To those who are opposed to IVF on religious grounds, the use of excess eggs from the procedure renders SCNT even more unacceptable.[7]

The Catholic Church does, however, support research using stem cells derived from adults or umbilical cords. The United States Catholic Conference of Bishops worked closely with legislators such as the conservative Senator Sam Brownback of Kansas in mustering support for legislation that would prohibit research involving any form of human cloning, and has actively campaigned for research employing adult or umbilical cord–blood stem cells.

Despite opposition from the Catholic Church and like-minded religious organizations, there has been a steady growth in support for legislation that would permit the federal government to fund research on stem cells derived from human embryos created in IVF procedures. The shift in support has come about largely because the public has increasingly come to believe that stem cell research has great therapeutic promise. The Stem Cell Research Enhancement Act of 2005 provided for an enlargement in the availability for research of stem cells derived from human embryos. In the face of vigorous opposition by the Catholic Church and closely aligned groups, such as the American Life League and the National Right to Life Committee, the act passed both houses of Congress in 2006. President George W. Bush, however, vetoed the bill, the first veto of his presidency. Bush has long held that terminating the viability of the embryo in culling the stem cells is tantamount to abortion, a position commonly put forth by right-to-life groups.

Stem cell research is controversial because some feel that it is fraught with ethical pitfalls—for example, those associated with obtaining the eggs needed for SCNT procedures. Second, and to the point of our present discussion, there are those who oppose on religious grounds all forms of such research, other than that carried out with adult stem cells or umbilical cord–blood stem cells, as outlined above. Opponents of stem cell research have availed themselves of every available recourse—including appeals to public opinion, political pressure, and the exercise of religious authority in the churches—in their efforts to curtail it. Science is thus faced with potent threats to its autonomy—namely, its prerogative for determining the limits of scientific research via mechanisms internal to science itself.

7. Interestingly, Francis Collins seems to find SCNT the least morally and ethically offensive of the various approaches yet formulated (2006, 252–57).

In some senses there is nothing new in this challenge; science faces limits to the scope of research all the time. It is understood that some kinds of research—for example, medical research that imperils human subjects or brings them evident harm—are out of bounds. Such constraints on research are accepted on general ethical grounds by virtually all scientists. Constraints on some research may also arise because there is no funding for particular areas of potential investigation. Among scientists it is understood that resource limitations do often arise, and some areas of study will come into or out of favor with funding agencies, both governmental and private. But the controversies over stem cell research are different. They pose more sharply and more visibly than any instance in recent times the question of who sets the limits on the range of permissible scientific study. A substantial fraction of scientists, including those most conversant with the technical details and ethical challenges, favors governmental permission for, and funding of, stem cell research that involves use of embryonic stem cells derived from IVF procedures, as well as therapeutic cloning. Their views are reflected in the policies advocated by scientific and medical associations such as the AAAS and the AMA (American Medical Association 2003). A committee of the NAS in 2005 issued guidelines for human embryonic stem cell research (NRC 2005). On the other side are arrayed those who would forbid or set limits to scientific study based on ethical and religious grounds not widely accepted in the scientific community.

The stem cell controversy is not just about laboratory-based science; it implicates biomedical applications that have the potential to cure or greatly alleviate a variety of diseases. Because this is so, there is a much broader awareness of the controversies, and a much deeper level of public participation in the discussions of what should be done, than is typical of issues that involve the interface of science with society. While initially the federal government was the primary locus of legislative action and funding, state governments have come to play increasingly important roles. Thus, for example, the state of California took a large step into this arena with passage of Proposition 71, a ten-year, three-billion-dollar bond measure to be used solely to fund medical research that will likely involve studies using SCNT. In the 2006 election, a referendum called the Missouri Stem Cell Research and Cures Initiative passed in the state of Missouri. Although it forbids all forms of human cloning, it assures that any stem cell research permitted under federal law can be pursued in Missouri. Other states have also passed legislation explicitly permitting stem cell research and addressing issues of regulation and funding. These developments demonstrate the extent to which stem cell research has become a matter of widespread civic concern.

The debates over stem cell research policy have most often been elevated to very public view by the participation of famous personalities such as Nancy Reagan, Michael J. Fox, and the late Christopher Reeve. These people have argued for stem cell research because it holds out the promise of ameliorating or curing Alzheimer's or Parkinson's disease, or reversing disabling spinal cord injury. Public figures, and the interest groups they represent, have often made their cases by conflating basic laboratory research with so-called translational research, which involves advancing basic laboratory results to the stage of human subject testing, and eventually to demonstrated efficacy and safety.

Stem cell researchers have allowed nonscientists to become the most visible advocates of stem cell research. By doing so they have in effect ceded some of their expert authority. The public takes the rosy predictions of medical advances, some of them quite unrealistic, as representative of scientific opinion. The British scientist Lord Robert Winston (2005) commented on the dangers for science in promising more than can reasonably be expected given our current understandings: "One of the problems is that in order to persuade the public that we must do this work, we often go rather too far in promising what we might achieve. This is a real issue for the scientists. I am not entirely convinced that embryonic stem cells will, in my lifetime, and possibly anybody's lifetime for that matter, be holding quite the promise that we desperately hope they will."

As of this writing, the prospects for stem cell research to yield beneficial therapies are clouded by several factors. Although it has great potential, human cloning via SCNT has not yet been achieved. The scandal created by the malfeasance of Dr. Woo-Suk Hwang and coworkers in South Korea has damaged the field's reputation. Nevertheless, attempts to successfully carry out SCNT have continued throughout the world. The Harvard Stem Cell Institute, in collaboration with researchers from Columbia University, announced in June 2006 that it would begin recruiting women in Boston to donate eggs in order to generate lines of embryonic stem cells. Spokespersons for the program made it clear that the research is very much in its infancy, and that clinical applications could be a decade or more away. The emphasis in the research will be on basic studies of cellular development. Thus, there is no prospect of early clinical application. Furthermore, progress in the research depends on the availability of eggs from the ovaries of female donors through a procedure that is potentially painful

and somewhat risky (because ovarian hyperstimulation can produce unto-
ward effects). It is not clear how clinical applications that might develop
could actually proceed, given prospective limitations on the availability of
donor eggs and significant ethical concerns relating to their procurement
and utilization (Magnus and Cho 2005).

As mentioned above, there are on the order of four hundred thousand
embryos, excess products of IVF procedures, stored at low temperature in
fertility clinics across the country; only some eleven thousand or so, however, are designated by their donors for use in research (Hoffman et al.
2003). The number of distinct new stem cell lines with differing genetic
properties that could be generated from this store would certainly be far
short of the very large number needed for widespread therapeutic use. In
that case, where might additional embryos be found? One possibility is
that they could be created solely for their use in therapeutic cloning. But
this would involve the same ethical issues identified above in connection
with SCNT procedures. In addition, it would highlight the entirely separate
ethical question of whether it is permissible to create human embryos
specifically for their use in research. This step would certainly be seen in
a much less favorable light than the use of embryos already created for use
in IVF procedures (and in all likelihood destined to be destroyed anyway).

New research may have found a way around this ethical issue by the
time that this book is in print. Preimplantation genetic diagnosis (PGD) is
often used to identify genetic defects in embryos created through in vitro
fertilization before they are implanted in the uterus. In this procedure,
one cell is removed from the blastocyst at the point at which there may be
eight pluripotent stem cells present. That one cell is then tested for genetic
defects. In a potential variant of the procedure, the cell that has been removed could be allowed to divide. One of the two cells could then be used
for testing, and the other cultured to provide a human embryonic stem cell
line (Klimanskaya et al. 2006; Simpson 2006). This proposal has not been
attempted with embryos that are to be implanted in a human uterus, but it
may in time provide a pathway through the ethical thickets that face the
field today; in the course of doing so, it could also provide a large number
of human embryonic stem cell lines. Many levels of permission on the part
of donors would be required, and the technique would not escape the pro-
scription of religious groups that oppose IVF procedures. Despite these obstacles, it would go some way toward allaying the ethical concerns of others.

In advocating for federal approval and funding for stem cell research,
the science establishment has consistently attempted to avoid confronting

religious authority on the question of whether human embryos represent human life with the same status as humans living in the world. It is not really a debatable topic from the perspective of a religious authority that defines the beginning of human life as the moment of conception. But this narrowly framed religious view stands against the widespread view of many that embryos do not enjoy such a status, and that research on embryos holds great potential for learning more about the nature of cellular development and the origins of diseases, as well as the longer-range goal of curing or alleviating many diseases. The struggle for public approbation has become a war of metaphors. By continually referring to blastocysts as cellular collections, and by emphasizing that only embryos successfully implanted in utero have the possibility of eventually becoming a fetus, science has attempted to counter the rhetoric of the religious right, which couches its discussions of human embryos in terms of "human life," "abortion," and "adoption" of embryos. Science's most powerful argument has been that stem cell research may hold the key to curing many feared and painful human diseases. In making its case, science must avoid promising more than it can deliver, in a time frame that meets the expectations of the public. At stake is its standing in society as a paragon of unbiased, disinterested inquiry and knowledge.

## WHAT PERSUADES?

The examples presented in this chapter of conflicts between religious and scientific authority show that the authorities of science and religion are of different kinds. They make contrasting and frequently irreconcilable demands on individuals and society as a whole. Nowhere is the conflict of authorities more evident than in the United States. In 2007 the Pew Center published results from the Pew Global Attitudes Project that showed an overall inverse relationship between "religiosity," as determined by several metrics, and per capita gross domestic product for forty-four countries. The United States is an outlier; religiosity is higher in relation to GDP than for most other nations. It is not surprising then that the separation of church and state is a locus for much of the conflict that arises between religious convictions and secular values grounded in scientific rationality. More is involved, however, than publicly visible contests such as those we have described; *individuals* must reconcile the competing claims of science and religion in judging many epistemic claims and forming moral and ethical judgments. Their struggles turn on more than simply questions of scientific literacy generally, or acquaintance with specific bodies of scientific knowledge.

Jürgen Habermas argues that for those with deeply held religious beliefs, the very act of attempting a "balance" between religious and secular convictions can jeopardize their self-assessment as pious persons: "A devout person pursues her daily rounds by *drawing* on belief. Put differently, true belief is not only a doctrine, believed content, but a source of energy that the person who has faith taps performatively, and thus nurtures his or her entire life" (Habermas 2008, chap. 5). It would thus seem that people with deeply held religious convictions are not persuaded simply by evidence and arguments grounded in scientific rationality when the matter at hand has the potential to shake the foundations of their beliefs. The engagement of science with peoples' thoughts, attitudes, and beliefs must occur at a broader level, one where scientific facts, methods of analysis, and ways of viewing the world are integral with the rest of culture. Winning over peoples' hearts and minds is not just about teaching science subjects in the schools; rather, it is about conveying a more holistic notion of how science works: its capabilities, limitations, and potential for doing good in society—about its capacity for moral action.

# SEVEN

## SCIENCE AND GOVERNMENT

This chapter is about the ways in which science exercises epistemic and moral authority in the affairs of government, and how governmental policies and actions in turn constrain the scope and autonomy of science. It is in some sense a continuation of the theme of chapter 3, which dealt with American science. Much has been written about the roles and status of science in the federal government over the past half century and more. I will not attempt to reprise this literature. Our concern is with the particular question of the limits of the exercise of authority by science in government, and on how science's authority, moral authority, and autonomy are circumscribed by politics. We begin with a brief introduction and recapitulation of themes already touched on, then follow with specific stories of recent vintage that reveal how government has given scope to science's authority, constrained it, or even trumped it entirely.

In 1978 Don K. Price wrote about the evolving character of the social contract between science and government in the decades following World War II and the implementation of Vannevar Bush's classic 1945 report, *Science: The Endless Frontier*. The scientific establishment traded effectively on its important contributions to the war effort in order to gain the federal government's acquiescence to support basic research as a foundation for practical progress and support of national military, social, and economic goals. The terms of the contract were a complex mix of reliance on procedures and authority structures that arose in the war effort, and the invention of new organizational structures that would assure governmental oversight and control of expenditures and policies. In the process, scientific research achieved a measure of autonomy, though of limited scope. In writing about this, Price speaks of "authority," but I believe he really meant what I have been referring to as autonomy: "The authority gained by the scientific community was not of the kind that could be defended as a matter of constitutional right; it was a delegated authority, and it depended on the continued confidence among elected politicians in the assumptions on which the tacit bargain was founded—that basic research would lead automatically to fruitful developments" (79–80).

Price identifies three types of governmental constraint on research: regulatory red tape, antielitism, and controls on the substance of research. The first of these is not central to our concerns here, although we will see that regulatory agencies can and do place constraints on science through rule making. Egalitarian, or antielitist, movements are a fixture in the competitions for resources. Those who are less successful in garnering federal support for their research efforts are tempted to claim that special interests have been able to set the terms of competition in ways that disadvantage the less well connected. But throughout this book, we have previously encountered another aspect of the antielitist attitude: science frequently produces results that contradict settled beliefs, upset traditional understandings, or have the potential to negatively affect particular economic interests. A commonly employed response is to invoke the image of intellectuals isolated in their ivory towers, indifferent to and largely ignorant of the practical world in which nonscientists live. To the extent that such sentiments are felt and acted on by politicians, antielitism becomes an element of political constraint on science.

Thomas Gieryn's account of the struggles of the social sciences to gain status and support from the federal government provides an interesting historical review of post–World War II science policy generally, and of the processes of boundary making within science and between science and society at large (Gieryn 1999, chap. 2). The social sciences had to struggle to gain recognition as legitimate science within the science establishment. Vannevar Bush was concerned mainly with securing autonomy for basic research in the natural sciences in university settings, free from the intense governmental oversight that had characterized wartime research efforts. He was not very interested in the social sciences, and made no effort to include them in his brief for a new approach to federal support of science. Second, the social sciences had to overcome popular notions shared by many congressional figures that they really amounted to no more than common sense, and hence were not deserving of the same sort of support received by the natural sciences. Eventually the social sciences were made a part of the NSF, but their position there remains a bit tenuous even now. The products of social science research sometimes touch public sensibilities in ways that excite strong reactions.

Price's third constraint on the substance of scientific research is central to our analysis of the place of science in government. A core element in the ethos of science is the position that science must be autonomous, free of constraints on what might be investigated or how research might be conducted. We have already noted that science is in fact constrained in many respects by moral and ethical codes common to society as a whole, such as limitations on the use of human and animal subjects or prohibitions on work that could wreak environmental damage or destroy property. The governmental constraints that concern us here are those that place limits on the topics that might be studied, or that limit funding for research in particular topic areas, for reasons based on ideology, to satisfy political or special economic interest groups, or to preclude anticipated applications of the basic science. Battles to preserve the autonomy of social science research need to be fought repeatedly. In addition, some proposals to limit the autonomy of science are motivated by the feeling that the discipline is excessively insular, or that too many scientists are insensitive to the perspectives of the larger society and unresponsive to the public's right to be both informed and involved in setting science policies and allocating resources.

In the passage quoted above, Price notes that the government has not deemed scientific research a constitutional right, but rather one delegated through the legislative process and subject to modifications through actions of the executive branch and regulatory agencies. But it is a fair question to ask whether a generally accepted right to unfettered intellectual pursuit should not include scientific research. The traditional view of science as supportive of a liberal society sustains the notion that democratic ends are best served by allowing science a wide scope for investigation. But modern science is not simply intellectual pursuit. It demands substantial resources, and only government—whether state or federal—has the resources to support it adequately over time. Further, because the outcomes of scientific research often occasion serious moral, ethical, social, and economic consequences, governmental oversight and regulation of research is inevitable. There is thus a tension between the notion of a "right to research," and the many forces brought to bear in controlling what scientists do and what uses are made of the products of research (Brown and Guston 2009).

We discussed in chapter 4 how the social structure of science and the internal dynamics that operate within it affect how science deals with various external pressures, particularly those emanating from government. Responses come largely from institutional science: professional associations such as the AAAS and the APS; the National Academies, which issue hundreds of reports each year under the aegis of the NRC; and public interest groups formed largely of scientists, such as the Union of Concerned Scientists, or Science in the Public Interest. When it comes to particular instances of governmental constraint on science, however, individual scientists often assume advocacy roles, though doing so may take them into the unfamiliar and often-unfriendly world of politics, contesting ideologies, or vested economic interests.

## THE POLITICS OF STEM CELL RESEARCH

We saw in chapter 6 that stem cell research has been an open field of conflict between science and religion. Scientific autonomy has stood in opposition to ethical and moral arguments, based largely on religious beliefs, for limiting such research. As noted there, the opposing forces seek to advance their positions through a variety of mechanisms, including public persuasion and political action. Here I want to look at two specific areas in which political forces have been important. Not surprisingly, it is generally the case that the degree of interest of elected officials in these issues, and the positions they hold publicly, are reflective of public attitudes. Whatever their origins, the ideological and religious positions held, or at least supported, by those with political power are consequential, and sometimes constitute formidable obstacles to an effective exercise of scientific epistemic and moral authority.

Our first topic for consideration is the contentious issue of what uses might be made of somatic cell nuclear transfer (SCNT). Recall that this technique involves removing the nucleus of a cell and implanting it in another cell that has had its original nucleus removed. It is commonly referred to as cloning, because the cells that form as the initial cell divides and multiplies are intended to possess the identical genetic information held in the cell from which the nucleus is taken. When the goal is to generate a group of stem cells for further use in research or therapeutic medicine, the process is referred to as therapeutic cloning. This is to be distinguished from reproductive cloning, where the aim is to produce an intact organism that is a genetic copy, as with Dolly the sheep. In August 2002 the American Bar Association (ABA) issued a resolution supporting limited therapeutic research using cloned human embryos. The resolution reads as follows: "RESOLVED, that the American Bar Association supports the freedom to pursue scientific knowledge for the improvement of human health and opposes governmental actions that would a) prohibit scientific research conducted for therapeutic purposes, including research involving cell nucleus transfer that is not intended to replicate a human being; or b) penalize individuals or research entities that participate in such research, provided that such research is conducted in conformity with accepted scientific research safeguards." The ABA action was prompted by the passage of a bill in the House of Representatives (HR 2505) and the introduction of a nearly identical bill in the Senate (S 790), the Human Cloning Prohibition Act of 2001. The Senate bill was introduced by the staunchly conservative Senator Sam Brownback of Kansas. The legislation prohibits any attempt to carry out human cloning for any purpose whatever, as well as shipment or receipt of an embryo produced by human cloning or any product derived from such an embryo. It calls for substantial criminal and civil penalties for those convicted of violating its provisions.

In its report, the ABA argues that "governmental action that would ban *all* forms of cloning, and thereby foreclose all potential avenues of medical advancement offered by therapeutic cloning, poses a direct and serious threat to freedom of scientific inquiry." The report presents an argument for constitutional protection of scientific inquiry, noting that prohibition of specific research endeavors historically has been extremely rare, and confined to state rather than federal laws. First Amendment protection of free speech is invoked to argue for free scientific inquiry, on the grounds that it is essential to the advancement of knowledge and the discovery of truth. In addition, the report finds protection in the Fourteenth Amendment's due process clause, in that the amendment "encompasses not only freedom from bodily restraint, but also the freedom to acquire useful knowledge."

The ABA resolution advocates the position that when proposed scientific work is entirely for beneficial purposes, and properly carried out, the government has no sound basis for its prohibition. The argument is couched in terms of the intellectual issues associated with freedom of inquiry, but also gains support from the recognition that there are "immense potential health benefits of therapeutic research." These are just the same grounds advanced by the science and medical communities (Rowley et al. 2002; AMA 2003).

Against this ameliorative view of scnt methods, we have the testimony of Leon Kass in 2001 before the House Committee on Health in its hearings on the Human Cloning and Prohibition Act of 2001. Kass, a professor in the Committee on Social Thought at the University of Chicago, was originally trained in medicine and biochemistry, but at the time of his testimony was best known as a bioethicist. In 2001, president George W. Bush named him chair of the President's Council on Bioethics, and he served in that role until 2005. He has written extensively on the ethics of human cloning (Kass and Wilson 1998; Kass 2002), and has spoke in favor of legislation prohibiting all forms of embryonic cloning, arguing that it will be virtually impossible to prevent widespread reproductive human cloning if therapeutic cloning is allowed:

> Once cloned human embryos are produced and available in laboratories and assisted-reproductive centers, it will be virtually impossible to control what is done with them. Biotechnical procedures and experiments take place in laboratories, hidden from public view, and for good commercial reasons these doings are concealed from the competition and everyone else. Huge stockpiles of cloned human embryos could thus be produced and bought and sold in the private sector without anyone knowing it.
>
> Worst of all, a ban on only reproductive cloning will turn out to be unenforceable. Should the illegal practice be detected, governmental attempts to enforce the reproductive ban would run into a swarm of practical and legal challenges, both to efforts aimed at preventing embryo transfer to the woman and—even worse—to efforts seeking to prevent birth after the transfer has occurred.
>
> . . . The only practically effective and legally sound approach is to block human cloning at the start, at the production of the embryonic clone. Such a ban is rightly characterized not as interference with reproductive freedom, nor even as unprecedented or dangerous interference with scientific inquiry, but as an attempt to prevent the unhealthy, unsavory, and unwelcome manufacture of and traffic in human clones. It would do what the American people want done: stop human cloning before it starts. (Kass 2001)

Kass goes on to criticize alternative legislation that would allow therapeutic human cloning while preventing reproductive human cloning. His arguments refer to the lack of specific provisions in the bill, which would constitute avenues for skirting around its prohibitions. Nowhere in his testimony, however, does Kass address the moral and ethical aspects of therapeutic cloning itself. He ends his testimony with these words: "The present danger posed by human cloning is, paradoxically, also a golden opportunity. The prospect of cloning, so repulsive to contemplate, is the occasion for deciding whether we shall be slaves of unregulated innovation and, ultimately, its artifacts, or whether we shall remain free human beings who guide our medical powers toward the enhancement of human dignity. The preservation of the humanity of the human future is now in our hands." In an address at the Duke University School of Law, Kass added a further elaboration on this theme: "But we should not be deceived: saying yes to creating cloned embryos, even for research, means saying yes, at least in principle, to an ever-expanding genetic mastery of one generation over the next" (Duke Law 2003).

Kass's writing displays a sweepingly cynical view of science. He sees it as secretive, at work in laboratories insulated from public view, responsible only to itself. Science in his view will all to readily fall prey to interests that would subvert for commercial gain any control procedures on therapeutic cloning. He also asserts that any form of cloning is inherently "unhealthy, unsavory, and unwelcome"—the very prospect of it is repulsive to contemplate. To permit therapeutic cloning would make us "slaves of unregulated innovation," not free human beings able to guide our medical powers toward the enhancement of human dignity. But Kass's slippery slope argument could credibly be invoked to call a halt to a great deal of scientific work generally seen as beneficial to society. We saw in chapter 5 that many substances produced as useful products can prove harmful in one way or another to some people. Many drugs produced in pharmaceutical research and eventually approved for use may at the same time prove toxic, addictive, or generative of horrendous side effects. OxyContin, a strong narcotic pain reliever that has turned into a deadly street drug, provides a recent example (Meier 2007; U.S. Food and Drug Administration 2001). Does this mean that drug research should be banned? Many scientific discoveries and their technological applications have the potential for tragic as well as beneficial use. Science and the rest of society have to work together to create policies and conditions that minimize the possibilities for harmful consequences of an otherwise beneficial new product or technology. From the perspective of regulating research, human cloning offers no more potential for disaster in this regard than other scientific frontiers. For Kass, however, the issue lies deeper.

Given that cloning of human embryos *for purposes of reproduction* is widely denounced in all quarters, and would be forbidden in any legislation that permitted therapeutic cloning, the prospects for such illicit activity would seem no greater if therapeutic cloning were permitted. On the other hand, much might be learned from the practice of therapeutic cloning that would advance any attempts at reproductive cloning. If one took a dim view of the capacity of science to police itself under strict guidelines and prohibitions, as Kass seems to, the threat of reproductive cloning would be increased by allowing therapeutic cloning. At the heart of Kass's case against therapeutic cloning, however, is simply the assertion that any form of human cloning is in itself an assault on human dignity. Kass's position in this respect aligns with that of conservative Christians.

The President's Council on Bioethics became a subject of controversy in March 2004, when two members of the Council, Elizabeth Blackburn and William May, were replaced by the White House at the end of their first two-year terms. Blackburn, a distinguished researcher in biochemistry and biophysics at the University of California, San Francisco, had been clear in her personal statement to the Council that therapeutic human cloning should be allowed to proceed (Brady 2007). Both Blackburn and Mays had been at odds with many other members of the Council. She expressed criticism of the group's finished reports, particularly a January 2004 report entitled *Monitoring Stem Cell Research* (Mooney 2005, chap. 12). The Council's three new appointees expressed reliably conservative views on research involving embryos, abortion, and related topics (Philipkoski 2003). Blackburn had strong things to say about the affair: "It has been the unspoken attitude of the scientific community that it is our duty to serve our government in this manner, independent of our personal political affiliations and those of the current administration. But something has changed. The healthy skepticism of scientists has turned to cynicism. There is a growing sense that scientific research—which, after all, is defined by the quest for truth—is being manipulated for political ends. There is evidence that such manipulation is being achieved through the stacking of the membership of advisory bodies and through the delay and misrepresentation of their reports" (2004, 1379).

The controversies over policies governing human cloning are only indirectly about the authority of science. The scientific basis for stem cell work is pretty much agreed on, even in the face of much uncertainty about the cellular processes that operate to give pluripotent cells their particular character. The effort to ban the creation of human embryos via SCNT is clearly an attempt to limit the autonomy of science. Although many individual scientists may agree that such a ban is called for, it is safe to say that a majority of scientists, and certainly the institutions that represent groups of scientists, feel otherwise. The legislative branch of the federal government is the avenue through which those who would ban human cloning have chosen to act. Although not all those advocating the ban are religiously motivated, it is nevertheless true that this is largely a struggle between science and religion, and that it is being played out in the political arena. Sam Brownback, the senator who has most vigorously pushed for the ban, is widely recognized as a pillar of the conservative right wing of the Republican Party, closely allied with Catholic interests.

Science can address the challenge to its autonomy posed by the proposed ban on human cloning by marshalling arguments based on freedom of inquiry, as part of a general assertion of the right to do research. But as noted earlier, the post–World War II history of science in the United States suggests that such a right is not generally conceded in American politics. Because science is politically, economically, and culturally so tightly woven into the fabric of society, and because it depends on society for nearly all its support, it is buffeted by many of the same social forces that limit and shape other sectors of American culture. Recognition of this reality can be seen in the strategies that science has employed in its opposition to the ban on all forms of human cloning. Primary among them is the argument that use of SCNT methods is an important component of a multifaceted program of developing treatments for various diseases, such as diabetes and Parkinson's. This aspect of stem cell research was highlighted during the run-up to the 2006 congressional elections, as recounted in chapter 6. Second, it has been repeatedly pointed out that because most other nations do not impose the limitations on stem cell research that exist in the United States, the nation risks losing research leadership to other countries, which means a potential loss of the most talented scientists, as well as valuable intellectual property rights. These arguments rely on expert authority, in the sense that science can aver that the potential benefits of such research are real; but in arguing directly that the work *should* be permitted, science is attempting to exercise *moral* authority.

Although there has been a consistently strong constituency for banning all forms of human cloning, opposition to the use of existing embryos left over from IVF procedures to develop new embryonic stem cell lines has declined over time as the claims for potential therapeutic benefits and substantial economic gains have captured the public's imagination. President

George W. Bush's stem cell research policy, announced in August 2001, allows federally funded stem cell researchers to work only on stem cell lines already existing at that time. The administration claimed that there were seventy-eight such lines, but only a small fraction of that number ever became available to researchers, and most of those proved to have limited utility. As mentioned in chapter 6, in early 2005 the House of Representatives approved HR 810, which would expand federal support for embryonic stem cell research. The measure, which explicitly would not provide funding to derive stem cells via SCNT, passed handily, by a measure of 238 to 194. Following the House action, Senate majority leader Bill Frist of Tennessee announced that, in a change of position on the issue, he would support the same bill in the Senate. Frist's move put him at odds with President Bush and brought swift and angry responses from Christian conservatives (Stolberg 2005). The bill did pass in the Senate in 2006, but fell to a veto by the president a few days later.

President Bush's veto of HR 810 illustrates the power of government to limit the autonomy of science by denying researchers access to federal resources needed to pursue a particular area of research. In vetoing the bill the president was remaining faithful to his commitment to the religious right. But the moral authority compass needle has been pointing increasingly away from him and toward science. The states and private sources of funding moved quickly to fill the lacuna created at the federal level by the president's veto (Holden 2006b). Although these sources cannot fully substitute for a lack of federal funding, they add to the moral pressure for a change in federal policy. Barack Obama's election as president in 2008 ensures that stem cell research will be supported in the executive branch during his administration (CNN Politics 2008; White House 2009).

My intent here has been to show how during a particular period of time, political considerations have affected an area of science fraught with ethical and moral questions, and how those considerations have affected science's freedom to perform research. The state of knowledge with respect to stem cell research and its ethical implications no doubt will have changed by the time this book is in the hands of readers. The potential of adult cells to yield pluripotent stem cells provides an example of how rapidly the field is developing, and how the nature of the debate is changed as a result. Research with adult stem cells is not burdened with the ethical concerns that apply to embryonic stem cells. To the extent that adult stem cells might serve instead of embryonic stem cells in research and therapy, the argument for using techniques such as SCNT is weakened. David A.

Prentice, formerly a professor at Indiana State University, and an employee of the Family Research Council, which advocates on behalf of those opposed to embryonic stem cell research, published in 2005 a claim that adult stem cell research has helped patients with at least sixty-five different human diseases. Senator Sam Brownback read into the Senate record the listing of some sixty-nine different human illnesses that he claimed were being treated by adult and umbilical cord–blood stem cells. Prentice's claim was seen as a challenge to the authority of established science, and three scientists contested Prentice's claims (Smith, Neaves, and Teitelbaum 2006; see Prentice and Tarne 2007 for his reply). Recent reports of successful conversion of human skin cells into embryonic stem cells (Yu et al. 2007; Takahashi et al. 2007) have opened up exciting possibilities for therapeutic uses based on adult cell sources. Nonetheless, big scientific obstacles stand in the way. All one can say at present is that the field is evolving very rapidly. In the course of that evolution it is likely that new ethical issues will surface, and with them new challenges to science's autonomy.

### STRATOSPHERIC OZONE DEPLETION

Stratospheric ozone depletion is a topic of great importance for the environment and human health. It is also a fascinating chemical story that begins with an exceptional man. Thomas C. Midgley Jr. was educated as a mechanical engineer at Cornell University. He graduated in 1911 and several years later joined the Dayton Engineering Laboratories Company, headed by Charles F. Kettering. He was put to work investigating the problem of "knock" or "pinging" in internal combustion engines. The problem of knocking in engines was quite serious; it essentially prevented running them under conditions in which they would be most efficient. During the five years Midgley spent on the project, he became a self-taught research chemist because he realized that the solution would be found in controlling the rate at which the fuel burned. Midgley and his research team eventually solved the problem with a gasoline additive composed mainly of tetraethyl lead. From a technical perspective his solution was very successful, and it ushered in the era of leaded gasoline.

General Motors absorbed Kettering's company, rebranding it as the Delco division. In 1928 Kettering asked Midgley to try to find a new refrigerant for use in home refrigerators manufactured by Frigidaire, another General Motors division. None of the substances then in use as refrigerants were suitable for home use; they were either toxic or flammable. By employing the same empirical methods that had led him to find tetraethyl lead as an antiknock agent, Midgley eventually came on an organic compound—containing just one atom of carbon, two of chlorine, and two of fluorine—that had all the right properties. It was a remarkable discovery; the compound was nontoxic, nonflammable, and had the right volatility to serve as a refrigerant gas. Midgley's discovery led to a joint effort with the DuPont chemical company to manufacture a series of substances that were all members of a family called chlorofluorocarbons, or CFCs, sold under the trade name Freon. Freon became the standard refrigerant for homes, as well as for many other uses. Air conditioning became an additional major area of application. Because the CFCs are of very low toxicity, they found a host of other uses, such as propellant gases in aerosol cans, foaming gases in the manufacture of plastic foam products, and covering gases in electronic manufacturing operations where an inert atmosphere was needed.[1] In short, Freon was big business.

Thomas Midgley's first discovery, the antiknock properties of tetraethyl lead, gave birth to a new industrial enterprise that eventually proved to be a major environmental calamity because of the lead's toxicity. His second great achievement seemed quite immune to similar unfortunate consequences. Because Freon was nonflammable and relatively nontoxic, it was approved by regulatory agencies for many applications in which it simply dispersed in the environment. Freon appeared to be an ideal product of human inventiveness; it was colorless, nearly odorless and tasteless, and did not give rise to any noxious products.

Scientists are naturally curious people. Someone was bound to ask at some point, Just what does happen to all the CFCs that find their way into the environment (Christie 2000)? Nearly all substances, including the thousands on thousands made by humans that are deployed in commerce, undergo some kind of chemical reaction in the environment that converts them into successor products, which may or may not be innocuous. They may be taken up by plants, consumed by bacteria, dissolved in the oceans and thence find their way into marine organisms, and so on. Scientists frequently talk about "lifetime"—the time it takes for the quantity of a substance released into the environment to dwindle to about 37 percent of its

1. Midgley became an eminent chemist, accorded many honors by the American Chemical Society. He was president of the Society at the time of his death. He was also elected to the National Academy of Sciences. Unfortunately, he contracted polio at the age of fifty-one and died four years later. See http://www.bookrags.com/Thomas_Midgley,_Jr.

initial value. For some very reactive substances, the lifetime is quite short, perhaps only seconds. For substances not chemically reactive, however, their lifetime can be quite long. For example, dioxins have a lifetime in the soil of ten to thirty years. The CFCS, which exist in the environment mainly as gases in the atmosphere (they are not very soluble in water), persist a long time, because there is no obvious chemical pathway by which they might be converted into something else. Professor Sherwood Rowland and Mario Molina, a postdoctoral researcher in Rowland's laboratory, undertook to learn how long the CFCS persisted, as well as the chemical details of their fates.

Molina and Rowland quickly came to agree with an earlier estimate due to James Lovelock that the lifetimes of the CFCS are indeed long, probably between forty and more than one hundred years. They were unable to identify any plausible chemical or biochemical pathways for removal from the lower atmosphere. One pathway that operates for many substances is decomposition due to the action of the sun's rays. This did not appear to be a possibility for the CFCS, however, because their decomposition by light requires radiation of high energy, which is not present near Earth's surface, where the CFCS are produced and released. The only pathway open for their decomposition thus requires that they diffuse up through the lower atmosphere and into the stratosphere, which begins roughly at the altitude at which long-distance jet aircraft fly. The stratosphere lies above the troposphere, where the weather we experience directly occurs, and where we humans live our lives.

It was well-known that the stratosphere acts as a shield to filter out high-energy solar radiation before it penetrates to the troposphere, and hence to Earth's surface. Indeed, the stratospheric filter is essential to the existence of life as we know it. Living materials could not very well withstand bombardment by the high-energy ultraviolet (UV) rays screened out in the stratosphere. A very important part of the filtering action is due to the presence of ozone, a chemical form of oxygen.[2] Ozone is a nasty chemical when present in the air of Los Angeles or any other city beset with smog problems. In the stratosphere, however, it is absolutely essential to a healthy planet. Ozone is produced in the stratosphere by the action of the sun on molecules of the air. At the same time, it is used up as it absorbs a certain band of high-energy solar rays. It is just this absorption of high-energy solar rays that makes ozone an effective filter. The concentration

2. For a brief, clear explanation of the importance of stratospheric ozone, see http://www.ozonelayer.noaa.gov/science/o3depletion.htm.

of ozone at any given time is the net result of processes that form it and others by which it is consumed.

From the perspective of public policy, Molina and Rowland's work might have been just another arcane study in atmospheric science. They saw, however, that CFC molecules diffusing up into the stratosphere encounter solar energy energetic enough to break the chemical bond between carbon and chlorine, thus liberating a highly reactive chlorine atom. At just this time Richard Stolarski and Ralph Cicerone proposed that chlorine atoms in the stratosphere would have the capacity to decompose an ozone molecule, and, by virtue of some chemical processes we need not go into, be regenerated to attack yet another ozone molecule, with the catalytic cycle being repeated many times. In this way a single chlorine atom could cause the decomposition of more than one hundred thousand ozone molecules before it becomes involved in some other atmospheric reaction. Taking into consideration that more than a million tons of CFCs were being produced annually, and that essentially all of that production would eventually end up in the atmosphere, Molina and Rowland (1974) warned that CFCs could pose a serious threat to the ozone layer. Their computer model predicted that ozone depletions of as much as 5 percent might already have occurred, and that eventually the depletion resulting from all the existing CFCs might be as much as 13 percent of natural levels. Ozone loss of this magnitude would allow a substantially higher intensity of UV rays to reach Earth's surface. It would thus pose increased risk of skin cancer for humans, and have the potential for damaging plants and oceanic phytoplankton as well. They urged that production of chlorofluorocarbons be halted until more detailed scientific studies yielded more information.[3]

The story that Molina and Rowland told about the potential effects of CFCs on the ozone layer was not supported by much direct evidence. To estimate the rates at which CFC molecules might diffuse up into the stratosphere, they relied on general arguments regarding the circulation of molecules in the troposphere. Measurements that would show that the chlorine present in the stratosphere originated in the CFCs and not in other natural sources were lacking. Neither was there at that time any significant evidence that the ozone layer was being depleted relative to historical

3. The story of how international agreements phasing out the production of CFCs and related halogen-containing compounds came into being is told in fascinating detail by Richard Elliot Benedick, a former career officer in the United States Foreign Service, who served as the chief U.S. negotiator in crafting the international agreements. He recounts the many twists and turns in both the U.S. domestic and international considerations of the issue (Benedeck 1991).

values. It would not be a simple matter to pin down a real decline in the ozone layer, because ozone levels are subject to many natural variations, such as an eleven-year cyclical variation in the intensity of solar radiation, volcanic eruptions, the effects of the El Niño atmospheric oscillation in the Pacific, and seasonal variations. In the absence of direct evidence of significant loss of stratospheric ozone, little progress was made in the effort to constrain CFC usage. The immediacy of the potential threat to human health and damage to plant life, however, helped galvanize environmental groups such as the National Resources Defense Council (Donige and Quibell 2007). In response, DuPont ran full-page newspaper advertisements in 1975 defending CFCs and questioning whether there was credible scientific evidence that there might be harmful effects from their continued use.

New scientific evidence showing that atomic chlorine was indeed present in the stratosphere began to accumulate. Congress responded to public pressure with an amendment to the Clean Air Act of 1977 that empowered the director of the Environmental Protection Agency (EPA) to regulate any substance with the potential to endanger public health by its effect on stratospheric ozone. The Food and Drug Administration, the Environmental Protection Agency, and the Consumer Product Safety Commission ordered the phaseout of "nonessential" uses of fluorocarbons in spray products, such as deodorants, hair sprays, household cleaners, and pesticides. There was considerable resistance from industry, however, to further phaseouts, which would include air conditioning and refrigeration applications, in part because suitable alternatives to the CFCs were not at hand. Indeed, industry mounted formidable public relations campaigns arguing against any further reductions in CFC production and use. Along with questioning the scientific arguments for potentially harmful effects, industry pointed to dire economic consequences of a phaseout of CFCs.

Molina and Rowland's discovery of the role that CFCs might play in depleting the stratospheric ozone layer exemplifies the exercise of both expert scientific authority and moral authority. These scientists did not begin their researches with the goal of demonstrating that CFCs are environmentally harmful. Having seen the potential for harmful effects, however, they were not content merely to report what would be likely to happen to the stratospheric ozone layer if CFC production went on increasing each year. They campaigned actively for a ban on CFC production, and sought every opportunity to make the scientific case for their position. In response, the parties with a vested interest in the sale and manufacture of the CFCs, and in applications employing CFCs, such as the aerosol can industry, sought to downplay the significance of the work and question the science. NAS estimates of the expected long-term depletion of the ozone layer varied in successive reports in the late 1970s and early 1980s, as scientific data accumulated and the theory used to make the predictions steadily improved. Those who wanted no action to be taken seized on the uncertainties that this variability revealed. Whereas antiregulatory interests and policy makers demanded authoritative answers to questions, free of ambiguity and uncertainty, scientists were able to offer something less solid: models that fit the data in hand, and that seemed to have predictive value. The inherent skepticism of scientists toward new results, and the deliberate, focused efforts to reduce uncertainties through new experimental studies, could appear to the nonscientist as waffling and a lack of direction for reaching consensus. Members of Congress pressured by special interests found it easy to disparage the scientific arguments: they were just theories, and were not founded on "sound science." Nonetheless, even though the ban on use of CFCs in aerosol cans went only a short way toward the eventual goal, it was a significant event: a new scientific idea with environmental import announced in 1974 had in just a few years gained enough traction to result in substantial government action. It was not inconsequential that DuPont, the largest U.S. manufacturer of CFCs, had put into motion a vigorous research effort to develop alternatives that would pass muster with respect to ozone depletion.

The net result of U.S. governmental action through the 1970s was thus a modest response by way of reducing the extent of CFC use. The effort to ban CFCs seemed to have lost its initial momentum, but the status quo was greatly disturbed in 1981 when ozone data from a British Antarctic station revealed a steep dip in ozone levels during September and October, the Antarctic springtime. A careful analysis of data from earlier years revealed that prior to 1975 the springtime levels had been more or less consistent, but from that time forward there was a remarkably steady decline. Despite the urgency felt by workers in the field, it took a few years of careful work and comparison with other measurements before it was quite clear that the effect was real. NASA scientists confirmed the British ground-based observations with satellite-based data showing that the effect extended generally over the Antarctic area. The story of the discovery and confirmation of the Antarctic "ozone hole," as it came to be called, illustrates how scientific results come to be thought of as "true." The process in this case began with an almost accidental notice of anomalous observations by the

British group. These measurements were not taken very seriously at first because they deviated so markedly from the expected trend. Steadily, with more data in hand and more detailed comparisons with other sources of data, and by successively eliminating potential sources of error, the effect was established as incontrovertible empirical evidence; there was indeed a seasonal Antarctic ozone depletion of major proportions. The metaphor of an ozone "hole" transferred readily into public discourse (Christie 2000, chap. 6).

The presence of the ozone hole was indeed alarming, but it had yet to be shown to have anything to do with human activities, such as the production of CFCs. It did, however, seem to be getting larger and more extensive each year. In 1986 a team of scientists led by Dr. Susan Solomon confirmed not only that the ozone levels dipped dramatically in the austral springtime, but also that chlorine was involved. Within another year or so, every other plausible explanation for the hole, other than that it was due to chlorine originating in CFCs, had been effectively disproved (Solomon 1988). Meanwhile, politicians had already begun to consider the possibilities for international controls over CFCs. As the evidence for a link between CFCs and ozone depletion gradually strengthened, scientific information on the problem was collected and distributed by the World Meteorological Organization (WMO) and the United Nations Environment Program (UNEP), with the aid of U.S. and West German science agencies and ministries. At the 1985 Vienna Convention for the Protection of the Ozone Layer, twenty nations and the European Commission signed an ozone convention. Remarkably, the United States and several allies introduced a resolution to authorize the UNEP to start negotiations on a legally binding protocol to be completed by 1987. In that year fifty-seven industrial nations signed the Montreal Protocol, setting in motion a phaseout of CFCs and other ozone-depleting chemicals. Evidence for an ever-increasing ozone hole continued to accumulate; in response, the internationally agreed-on timetables for phaseouts of CFCs and other ozone-depleting chemicals were accelerated.

As already discussed, a vital factor in establishing scientific authority relating to a contentious issue is an apparent consensus within the scientific community. Because ozone depletion is a global problem, a consensus of international scope on the scientific issues was essential. During the 1980s and 1990s, regular reports issued jointly by the WMO, a United Nations entity, and the UNEP were very influential in shaping public opinion and in promoting activity on the diplomatic front. The WMO/UNEP scientific assessment of 1991, compiled by hundreds of scientists from around the world, reported deterioration of the ozone layer far more severe than had been anticipated. The report included assessments of threats to human health, agriculture, and oceanic phytoplankton. Although there continued to be voices on the fringes arguing against the causal connection with CFCS and other human-generated halogenated chemicals, it appeared to nearly everyone that the scientific community was firm in its assessments, and that the predictions of future developments needed to be taken seriously. It was shortly after the 1991 assessment that DuPont announced that it would cease production of CFCS by 1996, and production of other implicated halogen compounds even sooner.

The scientists who had begun the hullabaloo, Mario Molina and Sherwood Rowland, had achieved a considerable fame both within and outside the science community. In 1995 they shared the Nobel Prize in Chemistry with Paul Crutzen, a European scientist who had initiated work on another potential pathway for depletion of the ozone layer (Rowland 1996; Molina 1996). In the statement accompanying the awards, the Royal Swedish Academy of Sciences noted that the scientists had contributed to staving off global environmental changes that could have had catastrophic consequences for human society.[4] The discovery of stratospheric ozone depletion and establishment of a causal link with chemicals of human manufacture affords a powerful example of the exercise of expert scientific authority *and* moral authority. Molina, Rowland, and Crutzen were the first to call attention to the potentially calamitous effects of practices widely integrated into commerce. Their initial research and conclusions needed to be validated by a huge data-gathering and analysis effort extending over years. Yet within two decades of Molina and Rowland's initial paper, the production of CFCS and several other halogenated chemicals capable of depleting the ozone layer had been greatly reduced through the agency of a worldwide treaty signed by more than 180 nations. At present, world production of CFCS hovers around 2 percent of peak production (which occurred in 1988).[5]

Successful though the process now appears, the course of events leading to the Montreal Protocol and its later modifications was difficult. At

---

4. All three scientists continue to work on environmental issues such as ozone depletion, urban air pollution, and global warming.

5. This assessment does not include CFCS illegally produced and sold (Benedeck 1991, chap. 17). Illegal activity has nearly disappeared, however, as refrigerant and air conditioning systems have been redesigned to accommodate CFC replacement substances, and old equipment has been updated or replaced.

many stages along the way it could have gone awry; early resistance came from industrial interests, but as the regulatory and diplomatic efforts moved forward, it came principally from antiregulatory quarters. I have not dwelt on these because the story is already told so well by Richard Benedick. It is worthwhile to note, however, that even as the science undergirding predictions of how much ozone depletion might occur and what effects it might have seemed more and more incontrovertible, those opposed to regulating production of CFCs and related compounds were unabashed about challenging the evidence. In 1995, when the Republicans swept into power in Washington promoting an avowedly anti-environmentalist agenda, members of Congress seeking to burnish their antiregulatory image challenged the scientific basis of the prevailing ozone-depletion model by invoking the opinions of antiestablishment scientists with questionable credentials and little or no research expertise in atmospheric sciences. Tom DeLay, one of the highest-ranking members of the House, introduced a bill to repeal U.S. implementation of the Montreal treaty. In the hearings that followed, it became evident that DeLay and others of his persuasion had not read the WMO/UNEP assessments (Benedick 1991, 226–28). Fortunately, industry had by this time invested heavily in alternatives to the CFCs, and was opposed to abandoning the course that had been set. Still, even in this apparently model example of a significant role for science in establishing regulatory and diplomatic policy, attempts were made to either question its authority or marginalize it in favor of other values.

Some have explained the success of the Montreal Protocol in terms of judgments of national self-interest and payoff structures (Sunstein 2007, chap. 2). In effect, the argument goes, the switch away from CFCs to alternatives, and the phaseouts of other halogen-containing substances, can be understood largely in economic terms: "Why did the United States adopt such an aggressive posture with respect to ozone depletion? A large part of the answer is that cost-benefit analyses, by the Council of Economic Advisers and others, suggested that the United States had far more to gain than to lose by a well-designed agreement" (Sunstein 2007, 82). Regardless of whether economic considerations were dominant, it remains true that science's expert and moral authority were central to the success of the effort to phase out CFCs and related chemicals. Without question, economic factors compete with scientific considerations relating to environmental consequences in all such regulatory and diplomatic decisions. To illustrate the sorts of compromises that are struck, consider methyl bromide. It is the major product used to fumigate soil before planting many fruit and vegetable crops, for postharvest storage and facility fumigation, and for government-required quarantine treatments. It is also a potent ozone-depleting substance. Participants in the Montreal Protocol agreed to phase out methyl bromide by 2005. The process began promisingly in this country, but in 2004 the EPA reversed the downward trend in U.S. usage by authorizing an increase in the use of methyl bromide, and permitting importation of large quantities. The argument given for the action is that methyl bromide is uniquely effective and suitable substitutes have not yet been found (U.S. Environmental Protection Agency 2008). It is argued that discontinuance of its use will cause considerable economic harm to many agricultural businesses. Because developing nations such as Mexico have a more relaxed schedule for phaseout than does the United States, the issue has become highly politicized. Environmental groups press for adherence to the Montreal agreements, and agricultural interests press for extensions. The debate is not about the science of methyl bromide's potential for depleting the ozone layer. Science has had its say; its expert authority on the matter is not in question. But with respect to how to act, science's voice is just one among several. This is how choices are made; when science has been heard from and believed with respect to how things are, what actions should follow are settled with deference to concerns that lie in other domains.

## GLOBAL WARMING

There is abundant evidence that Earth's climate has undergone many changes over time. The historical record, as compiled from observations of sediments, shows that much of the earth's landmass in the Northern Hemisphere has been repeatedly covered with glaciers. The glaciers appear to have come and gone at intervals of about one hundred thousand years. The most recent glacial episode peaked about eighteen thousand years ago, and then ended rather abruptly about ten thousand years ago. It is during the current interglacial period that humankind has prospered, multiplied, and colonized much of the planet's available land mass. The world's human population is estimated to be between six and seven billion. Demographers estimate that it might level out around 2050 at something like thirteen billion. The current population already has begun to place severe strains on the planet's resources, particularly in terms of food and energy (Friedman 2008, chap. 3). It is not at all clear how it will be possible to support twice the current population at an acceptable standard of living in the face of continually rising expectations.

Because humans are already crowding the planet and placing great stresses on its resources, any climate change that substantially altered the conditions of life on Earth could have disastrous consequences for many people. Just which groups would be affected, and to what extent, would depend on the nature of the change. Although a new ice age might develop on a timescale of perhaps a thousand years or less, it does not seem imminent. The more immediate prospect is that the planet will grow significantly warmer, and that substantial change will occur over a period of less than a century. Of special significance is that the prospective global warming appears to be of humanity's own making. So much has been written about this subject that there is no need to rehearse the scientific background in detail (Weart 2003). We must say enough, however, to see how the scientific understanding of climate change has arisen and gained currency in the larger society. Dessler and Parson (2006, chap. 3) provide an extensive overview.

Human activities over the past century or so, which continue to this day, have resulted in emissions to the atmosphere of certain gases that have the capacity to cause atmospheric warming. These so-called greenhouse gases include carbon dioxide, methane, and CFCs and other halogen-containing substances. The most important of the greenhouse gases is carbon dioxide, which occurs naturally in the atmosphere. The preindustrial level of carbon dioxide was about 280 parts per million (a way of expressing its concentration); in 2009 the level of carbon dioxide in the atmosphere will be over 390 parts per million. The large increase has arisen because of the burning of fossil fuels derived from ancient plant matter, primarily coal, oil, and natural gas. Combustion of wood and other plant matter of recent origin also produces carbon dioxide. One of the earliest scientists to recognize the ability of atmospheric substances such as carbon dioxide to act as a kind of blanket, helping to keep the earth's surface warm, was the British scientist John Tyndall, whom we met in chapter 5. The Swedish scientist Svante Arrhenius at the end of the nineteenth century first raised the prospect that the burning of fossil fuels could cause a global warming. Neither he nor anyone else, however, expected it to happen any time soon.

Since 1958, accurate measurements of carbon dioxide concentrations in the atmosphere have been made at Mauna Loa Observatory on a remote mountain location in Hawaii. These data show a steady increase, year by year, in the annually averaged carbon dioxide concentration. To interpret what this increase might mean for the global climate, we need to know what carbon dioxide levels were in past ages. Ice cores from the Antarctic ice cap, which go back 650,000 years, contain tiny air bubbles that can provide information not only about past carbon dioxide concentrations, but also about past average temperatures. There is a good correlation between the carbon dioxide concentrations and climate changes that occurred over that time (Siegenthaler et al. 2005; E. Brook 2005). During times of glaciation, the carbon dioxide levels were at their lowest; during warm periods, they were at their highest. The Antarctic ice core work provided important new evidence that went a long way toward convincing climate scientists that there is indeed a relationship between atmospheric carbon dioxide levels and climate. It is noteworthy that the level of carbon dioxide we experience today is the highest in that entire 650,000-year period. The question that needs answering is, To what extent could anthropogenic (of human origin) additions of carbon dioxide to the atmosphere, up to now and in the future, change climate by warming the planet? To even begin to answer this question, it was necessary to have a reasonably realistic mathematical model of the atmosphere, including its interactions with the solid earth, the oceans, and space. A sufficiently realistic model would need to be very complex, taking into account many interacting influences, which would mean stretching the capacities of computers to their limits.

One of those who dedicated his scientific career to the development of such a model was Syukuro Manabe, a graduate of Tokyo University who came to the Weather Bureau in Washington, D.C., in 1948. By 1967 Manabe and a collaborator, Richard Wetherald, working at the Geophysical Fluid Dynamics Laboratory in Princeton, New Jersey, had developed a global climate model (GCM) as sophisticated as possible in light of the computing power then available (Manabe and Wetherald 1967). Their model predicted that a doubling of the concentration of atmospheric carbon dioxide would lead to an average increase in global temperature of about 3.5 degrees Celsius (6.3 degrees Fahrenheit). This was an alarming result. The measurements of carbon dioxide increases year on year suggested that if they continued along the same pathway, the levels of carbon dioxide could be doubled by about 2050. Admittedly, the model omitted or oversimplified certain important factors, but it served as a convincing warning sign to many climatologists (Weart 2003, 112).

The global climate is a very complex system, affected by a host of variables, such as variations in the sun's intensity, the amount of soot and other particulate matter in the air (and at what elevation and latitude), the directions and magnitudes of oceanic currents, the amount of ice cover, and the rate at which carbon dioxide added to the atmosphere dissolves in the oceans—all of which are difficult to include in GCMs. Thus, while the Manabe-Wetherald GCM, and other GCMs that soon followed, all pointed in the direction of increased warming, the amount of such warming that might result from a doubling of the atmospheric carbon dioxide concentration was quite uncertain. In the intervening years, right up to the present, GCMs have grown in complexity and sophistication in pace with tremendous increases in available computing power. More and more factors have been included, and more of the uncertainties have been removed (Reichler and Kim 2008). The results of the model calculations have tended to be about the same: an increase in the concentration of atmospheric carbon dioxide from its current value of about 390 parts per million to about 850 parts per million by 2100 will result in an increase of average global temperature in the range of 1.7 to 4.4 degrees Celsius (3.1 to 7.9 degrees Fahrenheit), and a rise in sea level of from 0.2 to 0.5 meters (8 to 20 inches) (IPCC 2007a). The community of climate researchers has arrived at a consensus that the levels of carbon dioxide expected in the future, assuming continuing increases in the burning of fossil fuels, would result in significant and possibly long-term changes of the global climate.[6]

What evidence exists that the increases in the levels of greenhouse gases seen to date have actually produced a change in global climate? The question is not readily answered, because the predicted magnitude of change is still pretty small, and could easily be confused with minor climatic shifts of the same or larger magnitude that have nothing to do with human activities, such as variation in the intensity of the sun's output, changes in cloud cover, slight shifts in Earth's orbital motion about the sun, or other natural processes that interact with one another to create climate changes that range over decades or even centuries. Climatologists attempt to take account of all the other factors that might be playing a part by study of the record of past climate, as revealed in sediments, ice cores, tree rings—wherever climate change might leave a visible signature. Correlations are important, but as we saw in chapter 5, establishing a causal relationship based on a correlation when several other factors could also be operative is neither easy nor readily convincing.

6. Models that extend the forecasts into the more distant future, say 2300, and assume that essentially all the oil and gas reserves will have been consumed, predict a much more severe warming, with concomitant disastrous changes in ocean levels as well as other adverse environmental effects (Bala et al. 2005).

In large measure because of decades of scientific study involving many disciplines within science, global warming has become one of the most salient environmental and scientific issues of our time. Human society seems able to materially change Earth's climate in ways that will affect the lives of hundreds of millions of people. As we have already seen, depletion of the ozone layer is an example of such a capacity. In that case, it proved possible to arrive at worldwide agreements that greatly diminished the potential threats. Many people hoped that the processes that led to the Montreal Protocol and its later amendments could be emulated in forging agreements to abate carbon dioxide emissions. Richard Benedick and others built on their diplomatic successes in dealing with the ozone depletion problem to urge formation of the Intergovernmental Panel on Climate Change (IPCC), which was created in 1988 (Benedick 1991, 320–30). The IPCC has since continued to function as an agency for bringing together scientists with a wide range of expertise for fact-finding, analysis, and debate. But agreements regarding abatement of greenhouse gases that hold promise in themselves to materially slow anthropogenic global warming have not yet materialized. We need to ask why this is so, and what role scientific authority and moral authority have played during the past two decades.

When negotiations about how to curb emissions of greenhouse gases got under way in preparation for the 1992 United Nations Conference on Environment and Development, the enormousness of the difficulties that stood in the way of effective abatement became more evident. The agreement that was signed in Rio de Janeiro and ratified by about 160 nations (including the United States) did not impose the specific controls on emissions of the kind found in the Montreal Protocols. Nevertheless, it did embody moral commitments to reduce greenhouse gas emissions, particularly carbon dioxide, and aimed at reducing emissions to 1990 levels by the year 2000. Further diplomatic negotiating took place over the next several years, leading up to the December 1997 conference in Kyoto. The final results of the Kyoto negotiations, the Kyoto Protocol, call for the industrialized nations to reduce their greenhouse gas emissions by about 5 percent from the 1990 levels by the year 2010. For those industrialized nations in which emissions have been steadily climbing, the required reduction is in fact a very demanding target. The terms of the protocol include emission credits that can be traded between nations. A nation that is likely not to meet its targeted emission reduction can purchase emission credits from another nation that will more than meet its quota. Notably, the Kyoto Protocol does not place any limits on greenhouse gas emissions on developing nations such as China and India, even though both of them are already heavy emitters. As of December 2006, some 169 countries and other governmental entities have ratified the agreement, a sufficient number so that the protocol has become binding on the signatories. Notably, the United States and Australia have not signed the agreement.

Such protocols stand to burden the U.S. economy disproportionately in comparison with most other countries of the world (Sunstein 2007, chap. 2). American coal and oil interests, the automobile industry, anti-environmentalist ideologues, and oil-rich nations such as Saudi Arabia and Kuwait have resisted every step taken toward establishing limits on greenhouse gas emissions. Even the limited steps taken by the George H. W. Bush administration leading up to the Rio accords were accompanied by protests from these quarters. Vast sums have been spent on well-organized campaigns to sow mistrust of the diplomatic process, malign the underlying science, and predict economic hardships stemming from any attempts to limit emissions. Our interest here is in seeing how, amid great controversy, and buffeted by many contending interests, science has fared in terms of establishing consensus on the open scientific issues, exercising expert authority on those issues outside the science community, and creating a space for scientific considerations in the political arena.

As we have seen repeatedly, the exercise of expert authority depends on the perception of a scientific consensus. When there is an impression that scientists are in significant disagreement on a scientific issue, expert authority wanes. During the 1980s and 1990s, the potentially adverse consequences of a global warming became a rallying theme for environmental groups, along with many scientists who worked in one or another aspect of climate science or were well-informed on the scientific issues. In addition to the many indications that the earth is warming, however, there were many open scientific questions. The sheer complexity of the global climate system made it difficult to comprehend the problems in their entirety, to see how one variable affected many others. As policy debates moved front and center, those advocating legislative and regulatory action to reduce carbon dioxide emissions tended to emphasize results that pointed toward large environmental effects, whereas their opponents pointed to inconsistent findings and inherent uncertainties.

Any comprehensive model for the global climate will necessarily invoke a plethora of other models concerned with particular parts of the whole. For example, the global climate depends critically on the surface temperatures of the oceans, which are the results of many factors, such as oceanic currents, absorption of solar radiation, and exchanges of gases and heat with the atmosphere. Each model that goes into making the total picture is a complex metaphoric structure (T. Brown 2003, chap. 9). The end result is even more complex, as it encompasses not only these individual models for various components, but also the interactions between them. Added to that are numerous difficulties in properly defining and measuring the many critical parameters. For example, it is no trivial matter to put a number to the surface temperature of the planet (T. Brown 2003, 163–65). It is thus understandable that scientists have not been able to offer a simple and readily understood picture of all that goes into evaluating the earth's climate. Further, because so many scientific disciplines are involved in studies that bear on global climate, there is a significant prospect that disagreements will arise as to how to interpret the influence of some particular element in the overall model.

For these reasons, it has not been easy over the past few decades for scientists to arrive at a consensus position on global warming, let alone to convincingly project that view in society. But aside from differences in viewpoints regarding the science itself, economic and political interests opposed to any governmental actions that might reduce carbon dioxide or other greenhouse gas emissions have actively sought to counter emerging scientific evidence. One such organization was the Global Climate Coalition, an industry-based group founded in 1989, which included among its membership Amoco, the American Forest and Paper Association, the American Petroleum Institute, Chevron, Chrysler, Cyprus AMAX Minerals, Exxon, Ford, General Motors, Shell, Texaco, and the United States Chamber of Commerce. During George W. Bush's administration, the government has sought to control the public debates by muzzling scientists employed in governmental agencies, altering the contents of governmental reports in various ways, conducting congressional hearings carefully orchestrated to provide a public forum for contrarians and legislators opposed to taking corrective or ameliorative actions, and harassing scientists who produce unwelcome scientific results.

Senator James Inhofe, Republican from Oklahoma, is recognized as very conservative in outlook. His political campaigns have been heavily supported by the oil and gas industries. Until the Republicans lost control of the Senate in the 2006 elections, Inhofe was chair of the important Senate Committee on Environment and Public Works. No friend of any moves to regulate industry in light of environmental concerns, Inhofe used the committee to give voice to representatives of the oil and gas interests and contrarians on the issue of climate change.

On the Senate floor on July 28, 2003, Inhofe argued against the Climate Stewardship Act, cosponsored by Senators John McCain and Joseph Lieberman. The legislation would have broken new ground by creating caps on greenhouse gas emissions. The McCain-Lieberman bill failed by a 55–43 vote. As scientific evidence of global warming accumulated and public concerns grew, Inhofe's rhetoric changed little. One of his consistent tactics has been to call into question the idea of a scientific consensus. In a Senate speech on January 4, 2005, Inhofe said:

> For these groups [environmental extremists] . . . man-induced global warming is an article of faith. Therefore, contending that its central tenets are flawed is, to them, heresy of the most despicable kind. Furthermore, scientists who challenge its tenets are attacked, sometimes personally, for blindly ignoring the so-called "scientific consensus." . . . It's not hard to read between the lines: "skeptic" and "out of the mainstream" are thinly veiled code phrases, meaning anyone who doubts alarmist orthodoxy is, in short, a quack. . . .
>
> Since my detailed climate change speech in 2003, the so-called "skeptics" continue to speak out. What they are saying, and what they are showing, is devastating to the alarmists. They have amassed additional scientific evidence convincingly refuting the alarmists' most cherished assumptions and beliefs. New evidence has emerged that further undermines their conclusions, most notably those of the UN's Intergovernmental Panel on Climate Change—one of the major pillars of authority cited by extremists and climate alarmists. (U.S. *Congressional Record*, 2005)

Having set the stage for allowing contrarians their day in court, Inhofe went on to recite a long list of contrarian positions related to global warming. Despite his energetic efforts to portray a lack of consensus among climate scientists on the major points of climate change, however, Inhofe has steadily lost ground to a growing body of scientific evidence and concomitant consensus that anthropogenic global warming is real. The world of science, the public, and most governments of the world have almost universally accepted the fourth report of the IPCC, released on February 2, 2007, which convincingly demolished what credibility remained in the claims of global warming skeptics.

Philip A. Cooney spent several years during the administration of George W. Bush as chief of staff of the White House Council on Environmental Quality. Cooney is a lawyer, and has an undergraduate degree in economics. Prior to joining the Bush administration he was a lobbyist with the American Petroleum Institute, with a special focus on climate issues. He left his White House position in June 2005 to take a position with ExxonMobil, the world's largest energy company, as well as an unremitting opponent of any measures designed to reduce carbon dioxide emissions. During his tenure on the White House staff, Cooney was actively involved in editing reports from government agencies that dealt with global warming. He was reported to have collaborated in this work with Myron Ebell of the Competitive Enterprise Institute, an antiregulatory think tank that opposes governmental controls on fuel emission standards, carbon dioxide emissions, and similar measures to improve energy efficiency. Rick S. Piltz, formerly a senior associate in the Climate Change Science Program (the office that coordinates government climate research), resigned his position in March 2005, and has since compiled a long list of interventions by political appointees in publications of scientific reports relating to climate (Government Accountability Project 2006). On June 5, 2005, the *New York Times*, using documents acquired through the Government Accountability Project, revealed examples of Cooney's revisions or deletions of scientifically significant materials from reports from the Climate Change Science Program (Revkin 2005). These revelations are particularly troubling because in 2003 the White House Council on Environmental Quality and the Office of Management and Budget made major edits to a section of an EPA *Draft Report on the Environment*, deleting a section that dealt with the potential effects of climate change on ecosystems and human health, and rewriting parts of the report dealing with climate change. The deletions and changes were defended on the grounds that climate change would be handled by White House's Climate Change Research Initiative (Stokstad 2003).

James E. Hansen, longtime director of NASA's Goddard Institute for Space Studies, is that agency's top climate scientist, and has collaborated extensively with other scientists in research on climate change (Hansen et al. 2005). Hanson has repeatedly complained about efforts at the agency to restrict his access to public media to talk about global warming, especially since 2005. In December 2005 he gave a speech at the annual meeting of the American Geophysical Society in which he said, among other things, that significant emissions cuts could be achieved with existing

technologies, particularly in the case of motor vehicles (Hansen 2005)
Following his talk, according to the *New York Times*, officials at NASA
headquarters "repeatedly phoned public affairs officers who relayed the
warning to Dr. Hansen that there would be 'dire consequences' if such
statements continued, those officers and Dr. Hansen said in interviews"
(Revkin 2006).

Hansen's presentation at the American Geophysical Union meeting
consisted mainly of a presentation of the then-current understandings of
global climate change resulting from greenhouse gas emissions. In closing, however, he went beyond the simple presentation of the science. Here
are a few quotes from his text:

> Industrial emissions of $CO_2$ are declining. The two problem areas are
> emissions from power plants and emissions from vehicles. Solu-
> tion in both cases depends critically on improved energy efficiency.
> Power plants, because we need to avoid building a fossil fuel power
> plant infrastructure unless and until $CO_2$ sequestration is a reality.
> For vehicles, efficiency is critical because of the rapidly growing
> global number of vehicles. It is false to say that hydrogen technology
> will solve this in the future. It takes energy to make hydrogen. Effi-
> ciency will always be needed. It could be achieved now if we seriously
> pursued it, and if it were pursued we could get onto an "alternative"
> scenario track.
>
> My education-outreach team of students and teachers showed
> that in the U.S., even though the number of vehicles on the road
> increases every year, we could get off the path of increasing emis-
> sions by accepting even the moderate recommendations of the NRC,
> which calls for phasing in improvements in efficiency that would
> amount to about 30% by 2030. This would be based on available
> technology and it gives the automakers ample time to phase it in. It
> does require that they stop opposing efficiency requirements in court
> and stop pressuring lawmakers to oppose efficiency. . . .
>
> The goal of keeping further global warming under 1°C requires two
> things: first, flattening out and then decreasing the rate of growth
> of $CO_2$ emissions, and second an absolute decrease in emissions of
> non-$CO_2$ climate forcings, particularly methane and carbon monox-
> ide, and therefore tropospheric ozone, and black carbon (soot) aero-
> sols. There are multiple reasons to do this, with benefits for developed
> and developing countries, and for the planet and future generations.

But these are not things that will simply happen because they make sense. There has to be leadership. (Hansen 2005)

In these remarks and others, Hansen builds on his position as an authority on climate to offer normative views of what can and should be done to reduce greenhouse gas emissions (Hansen 2007). The Bush White House may not have wished to hear Hansen speak with clarity and forcefulness about the powerful scientific consensus on the prospects for global warming, but it most definitely did not want to witness him exercising moral authority, arguing for policy-laden actions that could be taken immediately to reduce the magnitude of undesirable climate change to come.

The Hansen case exemplifies tensions over what constraints on the exercise of expert or moral authority should exist for a scientist who works for the government, or for a corporation for that matter. On March 29, 2007, the Commerce Department posted a new administrative order governing "Public Communications." The order, which covers the National Oceanic and Atmospheric Administration, along with the Weather Service, requires scientists in the affected agencies to obtain preapproval to speak or write, whether on or off duty, on any scientific topic deemed to be of "official interest" (U.S. Commerce Department 2007). The policy excludes "Fundamental Research Communications," which have their own set of procedural rules, but applies to speeches, interviews, and more general writings in which the scientist might express a personal opinion.

Rep. Joseph Barton (R-Tex.) is an avowed skeptic on global warming. As chair of the House Committee on Energy and Commerce, with primary responsibility over the energy sector, Barton consistently acted over the years to prevent congressional action on global warming. In 2005, pursuant to proposed hearings on climate change, Barton wrote to several people demanding information. The recipients included three academic scientists, Michael E. Mann, Malcolm K. Hughes, and Raymond S. Bradley, who had collaborated on an analysis of proxy data for global temperature extending over the past millennium. The conclusion of the study was that Earth's temperature had in recent decades risen quite steeply, almost certainly because of anthropogenic factors (Mann, Bradley, and Hughes 1998, 1999). Barton's letter asked for a long list of information, including disclosure of all funding sources, any agreements made related to financial support of the research, the exact computer codes used in the work, responses to all referenced criticisms of the work, and the results of all temperature reconstructions carried out in the course of the research. In a editorial in *Science* critical of Barton's tactics, Donald Kennedy wrote, "It's clear that what's going on here is harassment: an attempt at intimidation, carried out under a jurisdiction so elastic that any future committee chair might try to play this game if coached by the right group of unschooled skeptics" (2005, 1301). Other scientific journals editorialized in the same vein (*Nature* 2005).

Michael Mann, in his written response to the Barton letter, addressed all the points raised. Barton's letter refers to criticisms of Mann's work, particularly in a paper published in the journal *Energy and Environment* by two Canadians, Steve McIntyre and Ross McKitrick. In reply, Mann pointed out that neither McIntyre or McKitrick is a scientist, to say nothing of having credentials dealing with any aspect of climate change. *Energy and Environment* is not a peer-reviewed science journal, and the journal editor acknowledged that the McIntyre and McKitrick paper was not reviewed. All of this background notwithstanding, Barton's committee held hearings on July 19 and 27, 2006, in which the research reported by Mann and his colleagues was critiqued in great detail (U.S. Congress 2006). Barton's heavy-handed attempts to call attention to the work of climate change skeptics and to cast doubt on the reliability of work such as Mann's brought forth a flood of reactions from scientific organizations and individual scientists. In response to the political pressures they experienced, Mann and others established a Web site at which their work and that of other scientists could be made available, along with links to other relevant sites (RealClimate 2005).

Working Group I of the IPCC is concerned with the physical science basis for understanding and predicting climate change. The group met in Paris in February 2007 to approve the long-awaited fourth assessment. The IPCC report is the product of the work of many smaller working groups of specialists within the climate change community. It must be approved by scientists from nations throughout the world. The fourth report builds on earlier work, but new observations and improvements in modeling have improved the reliability of the assessments of projected temperature changes, changes in sea levels, and other parameters related to climate change. It leaves little room for doubt that anthropogenic factors are responsible for significant warming of the planet, and that continued emissions of greenhouse gases will enlarge climatic changes (IPCC 2007b). We can be confident that there will continue to be resistance to the conclusions based on the report's underlying scientific studies. At the same time, the report's projections are inherently conservative, because they reflect the interests of many parties and governments, some of whom would like to see low projections for climate change. In light of observations made in the past few years, the estimates of sea-level change now seem to many scientists to be too low (Phillips 2007).

The improved extent and reliability of the data relating to many aspects of climate change, the energetic efforts of such organizations as the Union of Concerned Scientists and the Pew Center on Global Climate Change, the popularity of the movie *An Inconvenient Truth*, and a greatly increased level of coverage by the mass media have prompted governmental and corporate action at various levels. The state of California, which on its own represents the world's eighth-largest economy in terms of GDP, in 2006 passed legislation that sets targets for reduction of greenhouse gas emissions to 1990s levels by 2020 (California Assembly 2006). These limits are to be achieved by adopting market-based compliance mechanisms. The bill authorizes a state board to adopt a schedule of fees to be paid by regulated sources of greenhouse gas emissions. By February 2007, thirteen other states had also passed legislation requiring emission reductions (Pew Center on Global Climate Change 2008a). In addition, many corporations have committed to working toward addressing climate change issues. The Global Climate Coalition, mentioned earlier as being opposed to any steps directed toward limiting fossil fuel emissions, began losing corporate members in about 1997, and disbanded altogether in 2001. The Pew Center's Business Environmental Leadership Council, consisting of forty-two members and including large corporations such as BP, Duke Energy, DuPont, Exelon, Royal Dutch Shell, and Toyota, have committed to reducing greenhouse gas emissions through a variety of mechanisms (Pew Center on Global Climate Change 2008b). The Climate Action Partnership, composed of ten corporations and four environmental groups, has proposed a program of lowering mandatory carbon dioxide emission caps, coupled with carbon dioxide emissions trading (United States Climate Action Partnership 2008). Several faith-based conservative evangelical groups have urged followers to consider global warming as a religious and moral issue (Goodstein 2006; Moyers 2008). These examples are not meant to suggest that in themselves they offer responses to the challenges that must be addressed if significant reductions in emissions, particularly of carbon dioxide, are to be achieved. They do, however, reflect a growing acknowledgment of scientific authority relating to global warming. They also demonstrate a recognition that the difficult social, political, and economic challenges posed by climate change cannot be successfully dodged by interfering with the orderly workings of science.

I posed the question earlier as to why the abatement of chlorofluoro-carbon gases seemed to proceed much more rapidly than legislation and practice with respect to greenhouse gases generally. The answer, in part, is that the science delineating the relationship between greenhouse gases and climate change is more complex, and the effects are not being experienced so immediately. Neither are they, even after years of study, predictable with great precision. The answer is also, in part, that the forces resisting large-scale actions toward mitigation of greenhouse gas effects have been power-ful. The Bush administration and both houses of Congress have resisted any measures that in the short term might be damaging to the U.S. econ-omy. For example, the Byrd-Hagel resolution, passed 95–0 in the Senate in July 1997, resolved that the United States will not be signatory to any protocol or other agreement that would mandate commitments to limit or reduce greenhouse gas emissions if it would result in serious harm to the economy of the United States—a fine example of bipartisan shortsighted-ness (U.S. Congress 1997).

At some point in the past few years there occurred what Malcolm Glad-well (2000) refers to as a "tipping point." Perhaps it was triggered by the image of a polar bear standing on a small piece of arctic ice. Perhaps it was the influence of writings that vividly brought home what global warming actually means, notably Elizabeth Kolbert's *Field Notes from a Catastrophe* (2006). In any case, it appears that after four IPCC reports over more than a decade, the expert scientific authority of science on the major features of global warming and its relationship to greenhouse gas emissions has become an accepted element of public culture. Terms such as "green" and "carbon footprint" appear everywhere. But scientists and scientific organi-zations have been moving beyond the exercise of expert authority and mak-ing vigorous attempts to exercise moral authority as well (AAAS 2008c). As the science has advanced, the focus has begun to move away from the physical science of climate change and toward its social aspects. Economic issues are at the forefront. It is all very well to talk about mitigation by using alternative nonpolluting energy sources and reducing energy con-sumption, but even assuming their technical feasibility, what will it cost to implement new measures? How will those costs be distributed among nations, and what forms of compliance monitoring are feasible?

Cass R. Sunstein has argued that the responses of nations to calls for actions addressing global issues are primarily based on perceptions of national economic self-interest. He points to the various considerations at work, including international equity, intergenerational equity, and the moral issue of corrective justice (Sunstein 2007, chap. 2). Wealthy nations have been the principal contributors to the increased levels of greenhouse gases to date. Partially in recognition of this fact, the Kyoto Protocol imposes no restrictions on greenhouse gas emissions from developing nations, including China and India (as mentioned above). That these nations are exempted even though they are rapidly becoming major polluters, however, rankles many in the developed nations, particularly in the United States.

Economic analyses of the relative benefits and costs associated with materially reducing greenhouse gas emissions have produced numbers varying over a wide range, because the assumptions adopted in the models vary widely. Sunstein (2007, 91) refers to an industry-sponsored study by Wharton Econometrics Forecasting Associates that projected very high costs of complying with the Kyoto Protocol. Other studies, however, suggest much lower costs (Barrett 1999; Barrett and Hoerner 2002; Nordhaus and Boyer 2003). But aside from purely economic factors, major technological barriers stand in the way of substantial reductions in greenhouse gas emissions. A report from the Electric Power Research Institute points out the difficulties that lie ahead in achieving substantial emissions reductions (Specker 2007). Although these issues lie outside our immediate concerns, science will play important continuing roles in discussions of what remedial actions to take, because many new technologies will need to be developed and undergo evaluations. Each will have its own set of costs and consequences for environmental change.

## GOVERNMENT VERSUS SCIENCE

Nothing in the natures of science and government entails a necessary conflict between the two domains. Indeed, a great many governmental operations, including regulation of our complex society, absolutely require a sound scientific basis. In a democratic community, scientific considerations are often introduced in support of one position or another in the give-and-take of contesting ideas and interest groups. This fact frequently requires scientists and their institutional representatives to defend the integrity of science, which depends so critically on its autonomy and capacity for creating the expert knowledge and moral authority that society expects of it. During recent decades, and particularly during the administration of George W. Bush, this has been a strenuous task, as we have seen from the examples given in this chapter.

Much more might be written to illustrate the many ways in which the federal government has acted in recent years to limit the influence of scientific findings in formulating regulatory law and in the operations of regulatory agencies such as the EPA. The Data Quality Act, sponsored by Rep. Jo Ann Emerson (R-Mo.), and enacted as a rider to an appropriations bill in 2001, has been employed by special interest groups to challenge new agency regulations and even research findings. The legislation requires that the director of the Office of Management and Budget "provide policy and procedural guidance to Federal agencies for ensuring and maximizing the quality, objectivity, utility, and integrity of information (including statistical information) disseminated by Federal agencies" (Federal Register 2002). As detailed by Chris Mooney (2005, chap. 8), the act has been used by businesses to challenge government reports on such varied topics as climate change and the effects of herbicides on amphibian endocrine systems and diet. In the name of promoting what antiregulatory forces call "sound science," the Data Quality Act has served to stall or nullify regulations (Kennedy 2007). Although the various risk assessment guidelines employed by the federal government under the aegis of the Data Quality Act have met with enthusiastic approval by industry groups, the scientific community has seen them as impairing the effective use of science in support of regulatory decision making. An NRC committee of experts recently concluded that the Bush administration's 2006 draft guidelines for risk assessment are "fundamentally flawed" and should be withdrawn (NRC 2007).

Anyone who would like to see a more authoritative and consistently influential role for science in the federal government has not had much cause for optimism in the past decade or more. The Bush administration has been dismissive of scientific advice in favor of other voices, such as religious and political conservatives and business interests. Prior federal administrations have also shown a willingness at times to ignore scientific findings when under pressure from powerful interests. None, however, has so transparently attempted to subvert the authority of science through a multitude of artifices, some of which I have described in this chapter. The general strategy has been to (a) call into question the existence of a scientific consensus on contested issues; (b) intimidate key individual scientists or groups of scientists whose results or opinions are considered hostile to administration positions; and (c) impose regulatory and procedural barriers to the implementation of rules and policies already in place. It is heartening to note that these measures have not had lasting effect. As the processes that characterize the practice of science move forward, uncertainties are resolved, controversies are settled, and the force of what Polanyi referred to as "scientific opinion" carries the day. But this does not happen automatically. Ultimately the arbiter of many contested issues is public opinion. We are witnessing the truth of this as people become increasingly aware of what science has to say about the threats posed by global warming.

# EIGHT

## SCIENCE AND THE PUBLIC

The scientific discoveries of the past few hundred years have brought with them new, radically different understandings of both the physical world and our human nature. Galileo added experimental studies of the moon, Jupiter, and comets to Copernicus's model, and in the process made the new heliocentric universe seem much more vast than the old one. Harvey revealed a beautifully complex system for circulation of the blood, and in doing so gave new impetus to the idea that many mysteries lay hidden from sight within the body. The complex chemical nature of the world came into focus through the work of Antoine Lavoisier, Humphrey Davies, and other chemists. Since those early days, science's many revelations about the physical world have in one way or another continued to present challenges to traditional ways of thinking. Along with a new sense of the world's beauty and complexity, science has also made nature seem less predictable, less settled, and at times more intimidating.

The growth of science has also had important practical consequences. In the course of the past two centuries people in many parts of the world have been largely freed from many frightening, deadly diseases and recurring epidemics. A host of products that make for a safer and more comfortable life have become available. Much labor has been elevated from a grinding, subsistence level to more efficient and rewarding work. The world has been remade into a global civilization driven by science and technology, and much good has come of the transformation. At the same time, it is easy to find examples in which the changes wrought have had undesirable effects.

We now live in an increasingly crowded world, in which many resources have become scarcer, and in which adverse environmental consequences of implementing new technologies are visible. Because the distinction between science and technology is lost on most nonscientists, deleterious effects of the applications of scientific discovery are associated in the public mind with science. Thus science has, in the course of its evolution in Western society, acquired a public reputation that is a mixture of admiration and respect coupled with a certain amount of fear and mistrust.

Most who work within the scientific community think of science as a disinterested, rational search for new knowledge subject to skeptical evaluation. This self-image leads rather easily to a certain overweening sense of pride. Many in the science and technology community are sure that scientific thinking is superior in important respects to other modes of thought. When certain applications of scientific discovery have been criticized, often by those outside science, the challenges have been vigorously opposed, for a variety of reasons. Elof A. Carlson has described some of these instances in his book *Times of Triumph, Times of Doubt* (2006). One of his examples is the unbridled use of pesticides and herbicides in the decades following World War II. Rachel Carson, in her book *Silent Spring* (1962), examined the deleterious effects of chemical pesticides such as DDT on plants, animals, and humans. It made people think about the environment in a way that they never had before. Carson, a marine biologist, and already a popular writer on the marine environment, showed that a new technology that seems harmless and beneficial might have serious long-term effects. The chemical industry that produced pesticides and herbicides, and the agricultural establishment that was heavily invested in their uses, were shocked by the criticism. Carson was vilified in much of the chemical business community, and portrayed as hysterical and unqualified to write on the subject (Lytle 2007; Murphy 2005). Despite the attacks, or perhaps because of them, the book was a runaway best seller. It can fairly be said to have launched the environmental movement, which exhibits an undercurrent of suspicion and distrust both of corporate America and of science that, through its application, opens the door to manifold environmental harms.

Science's position as a source of expert authority, and especially of moral authority, is no sinecure. The public may respect science for the good that has come in its name, but it has also learned that its connections with other societal sectors, such as government and business, can lead to damaging applications of scientific discovery. We begin an examination of the public's views of science with a brief look at a particularly interesting episode in the 1970s that resonates even today in discussions of genetically modified foods and the implications of genetic medicine.

### RECOMBINANT DNA

Recombinant DNA (rDNA) methods are today the most indispensable set of tools in genetic research. Technologies based on them have led to many products related to treatment of disease, vaccination, diagnostic tests for disease, and genetically modified crops and foods. Examples include the production of human insulin used in treating diabetes, hepatitis B vaccine, and human growth hormone. But this powerful tool, borne of basic research in life sciences, had a tempestuous beginning (Krimsky 1982; Grobstein 1979). Methods related to rDNA were invented in the early 1970s; Paul Berg, a leader in this field, received the Nobel Prize in 1980 for his work. In a water medium, long DNA molecules from bacterial cells are cut into pieces in a controlled way—that is, by snipping them at precisely known locations with chemicals. The sections of DNA from one organism are then spliced back together with sections from another species to form a new DNA molecule. In this way one could insert the gene for making a particular protein, for example, into a bacterial DNA molecule. When the modified DNA is introduced into bacterial host cells, the protein is then formed in the bacterial culture.

In the earliest days of recombinant DNA research, the scientists who worked with the method were concerned about possible adverse consequences. The science was very new, and there were many uncertainties in how the biological systems under study would behave. In the words of the scientists themselves, the possibility existed that rDNA experiments could "result in the creation of novel types of infectious DNA elements whose biological properties cannot be predicted in advance" (Berg et al. 1974, 303). The scientists worried also that the techniques could increase the virulence of viruses or the resistance of bacteria to treatment with antibiotics. The fear that gene splicing could produce epidemic pathogens was heightened by the fact that biologists were using microorganisms in their rDNA research that have human hosts, most notably the bacterium *E. coli*.

In the summer of 1973, several leading molecular biologists wrote to the NAS about their concerns. In response, a Committee on Recombinant DNA Molecules was appointed to act on behalf of the Academy. The group published a letter in the journal *Science* proposing a worldwide moratorium on certain types of experiments as a prudent precaution until more was known (Berg et al. 1974). Not long afterward, at the famous Asilomar Conference in February 1975, more than 140 molecular biologists and geneticists from around the world, under the leadership of Paul Berg and others, endorsed a set of guidelines for rDNA research (Berg et al. 1975). This was an auspicious example of scientific self-governance. Instead of asking simply, "What can we do?" these scientists asked, "What ought we do?" Those at the forefront of an area of research recognized the need for a careful approach to a new and largely unknown terrain, and organized themselves to take precautions. Further, they exercised considerable moral authority in urging all members of the relevant scientific community to do likewise.

It was, of course, not sufficient by itself. There needed to be regulations in the form of detailed guidelines with respect to containment facilities and safety procedures. What followed over the next few years can fairly be described as chaotic. The task of formulating the research guidelines fell to the NIH. Leading molecular life scientists called on the agency to put into place a set of containment guidelines so that researchers around the country could adhere to a common standard. But the scientific community itself held widely divergent opinions on what constituted safe procedures and adequate containment. A DNA Regulatory Advisory Committee, established under the leadership of Donald Frederickson, director of the NIH, in due course produced a set of guidelines for rDNA experiments.

In the meantime, controversy over rDNA research had spilled into the public sphere, and citizens, environmental groups, local news media, and politicians all got into the act (Frederickson 2001). In Cambridge, Massachusetts, citizen groups, environmental organizations, and local elected officials engaged in spirited debates at city council meetings about the kinds of rDNA research that should be allowed within the city's boundaries. A particular issue was whether the city should approve construction of a "P3 laboratory" (the second-most restrictive kind) at Harvard University. The fact that a laboratory featuring extensive safety features was being planned suggested that the research would involve work with dangerous pathogens. The Cambridge Experimentation Review Board was established to review and approve policies in connection with rDNA research, and local ordinances to regulate the research were adopted. The Board conducted site visits and reviewed containment measures for all proposed experiments, in the interest of protecting residents from potential health risks. Nonspecialist laypeople played important roles in all these activities and in formation of policies. At a critical meeting in 1977, when the Cambridge City Council was considering extending a moratorium on rDNA experiments that posed medium to high risk, Maxine Singer, a distinguished biochemist and geneticist, who had cosigned the letter in *Science* that first raised concerns about rDNA research, was there to present the newly released NIH guidelines. The members of the City Council were meant to be reassured by the fact that a distinguished scientist with an international reputation had come to Cambridge to present and explain the significance of the new guidelines that would govern the conduct of rDNA research. It was clear sign that the scientific community took seriously the concerns of the public and was intent on addressing them.

The tumult in Cambridge was seen also in Ann Arbor, Michigan; San Diego, California; and other towns that were home to large research universities. Regulation of rDNA research became a national issue. Between 1976 and 1979 no fewer than twelve bills were introduced in the U.S. Congress that dealt with regulation of rDNA research. Some favored strict regulations with severe penalties for violations; others were more trusting of scientists to regulate their own work. There was such a range of opinion on the matter that, in the end, no bill was passed. Over time, however, the tension and fears subsided, particularly as further experience indicated that no harmful side effects were appearing from experimental work with the modified *E. coli* strains being used. Nonscientists were eventually appointed to the NIH's Recombinant DNA Advisory Committee, a move that helped to reduce public concerns regarding openness of the Committee's proceedings. The NIH guidelines remain in effect today, modified over time in the light of experience, and no new pathogenic organisms have inadvertently been released to the environment in the course of three decades of rDNA research.[1]

This story of the introduction of rDNA research into practice, and its eventual acceptance by all but a small minority of the public, illustrates that scientists can be keenly observant of their moral and ethical responsibilities not only to others in the scientific community, but also to society at large. It also illuminates the importance of communication between the scientific community and the public. In the face of widespread fear among the citizenry that harmful products might result from a new scientific methodology, research in the area was subjected to extraordinary scrutiny. It might have been brought to a halt entirely had not acceptable rules of

---

1. Ira Carmen (1985) has written a fascinating account of the rDNA debate in the light of constitutional issues relating to the right to conduct research.

practice and various monitoring procedures been implemented. Participation by representatives of the public in establishing those standards of practice and oversight was crucial. Also key was the active participation of scientists in public discussions, and in shaping the coverage that the controversy received in the media (Altimore 1982). The rDNA story reveals some of the many ways in which science and the public influence each other. To further explore this relationships, however, we must start with some analysis of what we mean by "the public."

## THE DIMENSIONS OF THE SCIENCE-PUBLIC NEXUS

A perennial theme of this book has been to examine the ways in which science exercises authority and moral authority in society, and how society grants a measure of autonomy to the scientific enterprise. Intrinsic to the discussions is the idea that the relationships between science and society are, in a sense, bilateral. We have seen, however, that science possesses characteristic social structures that promote insularity and impede open communication with the larger society. Most research scientists do not consider their work to be miscible with the affairs of society in the ways that a social worker, investment adviser, newspaper reporter, or teacher would. Consistent with that, science's image as a social entity in the larger society is that of a community more evidently bounded than many others. Among the features that distinguish it from other societal sectors is the sense that science is less transparent, less knowable by those outside it (Gieryn 1999, 345). Thus there is a special resonance to the idea of a "territory" of science. There are, of course, important interactions of many kinds between science and society, and we have discussed several of them in prior chapters. It is precisely because these interactions do exist that it makes sense, as Gieryn argues, to talk of a dynamic process of boundary making.

The enormous success of science, and its powerful influences in shaping the way that humankind understands itself and its place in the physical world, ensures that nearly every sector of society is inextricably linked to some aspect or another of science and technology. It is thus impossible for science to exist in isolation from the many forces that contend for power and influence in society at large. Science has enormous cultural authority of a certain kind, as discussed at various points in this book. Nearly every person, even one whose knowledge and understanding of science is small, comes into contact with the results of scientific discovery. Think, for example, of people sitting before a television set, availing themselves of emergency room care in a large city hospital, or receiving a genetically engineered medication to treat a blood disorder. They may not have an understanding of the science that underlies the technologies before them, nor of the pathway from basic scientific discovery to technological application, but their lives are nonetheless shaped by those products, and, ipso facto, by basic science. Even for the least educated, it is inescapably evident that something out there has made the modern world what it is. Their understanding of that something, and their attitudes toward it, help to determine the place of science in society.

Before we can even begin to ask questions regarding the authority relations between science and the public, several preliminary issues must be settled, among them:

- What do we mean by "the public"?
- What contacts exist between science and the public?
- What is the level of scientific literacy among the public?
- What is the interest level and attitude of the public toward science?
- What are the public's sources of information about science?

With respect to the first question, "the public" clearly refers to some kind of collective entity. Each person in society has a particular outlook on social questions. By "public" we refer to a group sufficiently diverse and large that its collective interests and concerns transcend narrow, more immediate personal interests of individuals or small groups within the larger sample. But in what sense can one express the attitudes and beliefs of a collective? In the simplest way of looking at it, the public is no more than the sum of a group of individuals. But each individual is made aware of the attitudes, thoughts, and arguments of others through various means, and is influenced by them to some degree. Richard Dawkins speculates that units of cultural information, which he calls memes, are spread from one person to another in society. They are, metaphorically speaking, infectious agents that propagate ideas, facts, and attitudes (Dawkins 1976, chap. 11; Blackmore 1999; Brodie 1995). Thus, for example, a person with little or no knowledge of baseball might hear in casual bar conversation that the Chicago Cubs are a lovable team, but that the Chicago fans know that they never win when the pressure is on. She might repeat that unit of cultural information, or meme, to others in another later conversation, and in this way contribute to the Cubs' whimsical, slightly poignant reputation. More to our point, public understanding of topics such as global warming and stem cell research is also, in part, the product of the diffusion of bits of information received by individuals in various contexts and passed on to others. The theory of memes has been criticized for a lack of testable hypotheses; its critics contend that it can be viewed as just another clever way to talk about processes that are vastly more complex than suggested by memetics. As a metaphorical model of the birth and career of bits of social knowledge, the notion of memes seems to me to have considerable merit. In any case, we know that individuals in society come by their knowledge of science and their attitudes toward science in general, as well as of specific topics, through a variety of channels, including direct interactions with others. Because so much of this is experienced in common, it makes sense to think in terms of a collective entity—the public.

The survey instrument is a dominant means of learning about public knowledge and attitudes. A properly formulated and executed survey can provide a kind of global view of what people know and feel about many matters. As an example, the 2001 Virginia Commonwealth University Life Sciences Survey consisted of a telephone survey of 1,122 adults nationwide. The respondents were asked more than fifty questions, with a provision for one of several responses. As a second example, the NSF regularly publishes *Science and Engineering Indicators*, which includes a chapter on "Public Attitudes and Understanding," which brings together and analyzes the results of various surveys (National Science Foundation 2008).

How does the public acquire scientific information? The Pew Internet and American Life Project recently reported the results of a survey of two thousand American adults aged eighteen and older, in which the respondents were asked where they got most of their news and information about science (Horrigan 2006). Television was identified by 41 percent of respondents as their primary source of science news; 20 percent said that they turn to the Internet. Newspapers and magazines each accounted for 14 percent of respondents. Interestingly, those with broadband capability at home were much more likely, and younger respondents were the most likely, to rely on the Internet. When it comes to a specific topic, the Internet serves as the primary resource for people. For example, two-thirds of those receiving questions about stem cell research said that they would turn to the Internet first for information on the topic; 11 percent said the library. The level of interest in news about science and technology was roughly equal to that for religion, national politics, and business and finance, and well below that for weather, crime, health, community news, and sports. Men are more likely to follow the latter topical areas, women the former.

Science is discussed in other media venues not normally considered sources of information about science. The rise of blogs on the Internet has opened up avenues for introducing topics with significant scientific content. While science-related blogs are a small part of the blogosphere, the most popular sites receive many visits (Butler 2006). In a related vein, many other Web-based initiatives are providing avenues in which science-related materials can be disseminated. Web 2.0 is a broad term referring to any media that involves the active involvement of the consumer, which includes sites such as Facebook, YouTube, MySpace, and Wikipedia. A danger associated with these media is that they offer boundless opportunities for dissemination of misinformation about scientific topics that it is virtually impossible to counter effectively.[2]

Talk radio represents a popular medium of communication that has a growing audience (Project for Excellence in Journalism 2007). Rush Limbaugh ranks at the top, but he is closely followed by several others who have gained substantial audience share. Six of the top seven talk radio personalities in terms of audience size in 2006 are avowedly conservative. The primary function of such programming is entertainment, structured around topics of social and political interest at the moment. When the talk is about something with scientific content, such as stem cell research, abortion, or global warming, talk show hosts do not generally take pains to present a balanced view of the topic. The mainstream scientific view is often ignored in favor of the contrarian views of an outsider, or the science is misrepresented by selective quotations. Unfortunately, talk show hosts represent virtually the only source of information and opinion related to science that reaches many Americans.

Few Americans are well-informed about new scientific discoveries and the uses of new inventions and technologies (National Science Foundation 2008, chap. 7). Surveys have also revealed that scientific literacy is quite low, not only in the United States, but in Europe and elsewhere. NSF surveys have attempted to assess general understanding of what it means to study something scientifically. The conclusion is that two-thirds of Americans do not have a firm grasp of what is meant by the scientific process. In a survey conducted in 2001, Americans and Europeans were asked thirteen

---

2. Wikipedia may be an exception to this generalization, because of its particular organizational structure, though it is inherently subject to at least short-term error content (Wikipedia 2008).

questions that tested their knowledge of science (Miller, Pardo, and Niwa 1997, 44). The questions (some in the form of true-or-false statements) included the following, along with the *percentage of correct answers* given by American respondents:

- How long does it take for the earth to go around the sun? (47%)
- Radioactive milk can be made safe by boiling it. (61%)
- The earliest humans lived at the same time as the dinosaurs. (48%)
- Electrons are smaller than atoms. (44%)
- Provide a correct open-ended definition of DNA (21%)

Scientific literacy can be assessed in terms of three components:

- A general knowledge of scientific facts.
- An understanding of scientific methodologies, of the characteristic ways in which scientists go about their work.
- An understanding of the nature of scientific institutions, of how science is actually practiced and of the social structures around which it is organized.

Based largely on the first two criteria, the researchers concluded that only about 20 percent of Americans can be said to possess civic scientific literacy—that is, sufficient knowledge of science and understanding of scientific practice to be able to grasp the essential scientific implications of matters that come before the public. The low literacy figure is echoed in more recent work. It seems possible, however, to love science even without knowing much about it. One interesting result from the Virginia Commonwealth University survey is that 86 percent of respondents judged that developments in science and technology have helped to make society better. An even higher 96 percent believe that it is very or somewhat important for the country to encourage more people to enter careers in science and technology. 72 percent of Americans agree that "the benefits of scientific research outweigh any harmful results."

Given the low level of general knowledge and understanding of science in the public, can there be effective public participation in matters dealing with science? We have seen in prior chapters that science often attempts to exercise authority in situations in which it is in conflict with alternative ways of evaluating a situation. We can hardly expect the public to competently evaluate scientific assertions in relation to competing claims if its understanding of science is very low. Even if there is a general inclination to accept scientific authority because there is a general belief that science conveys social benefits, how does the public distinguish genuine science from marginal work or quackery? A survey conducted in 2001 showed an increase during the prior decade in belief in paranormal phenomena such as extrasensory perception (abut 50 percent), or that people can hear from or communicate mentally with someone who has died (28 percent) (Newport and Strausberg 2001). Can scientific authority trump such beliefs? It would seem that they fall within a realm of emotional and religious commitments that simply survive rational attempts at dislodgement.

### ETHICS IN SCIENCE

While a substantial fraction of Americans have a supportive attitude toward science, they express concerns about moral issues. Some 72 percent either strongly or somewhat agree with the statement, "Scientific research these days doesn't pay enough attention to moral values" (Virginia Commonwealth University 2001). This view is held most strongly by those who report themselves as "religious," and least strongly by those who regard themselves as "very informed on science," but even in the latter group 55 percent strongly or somewhat agree. The public's perceptions of moral and ethical issues in science are clearly of importance in determining its assessments of not only science's moral authority, but its expert authority as well.

Moral and ethical concerns about science can arise at various levels: with respect to science generally, or with respect to particular areas of science, particular institutions, or the work of individual research groups. At the most general level, feelings of dissatisfaction arise from those who believe that it is not enough for scientists to apply the most obvious ethical standards, such as humane treatment of animals, doing no harm to human subjects or the environment, and following rules of good conduct with respect to truthful reporting of results. They call for a more comprehensive set of guidelines, which look beyond the immediate production of research results to concerns for how the newly gained knowledge will or could be used, or for what consequences might arise inadvertently. We will discuss the basis of such arguments later. Here I want to identify a broad range of ethical concerns, ranging from the misconduct of individual researchers to institutional policies, that weaken the public's trust in science.

Research Misconduct

Misconduct by individuals related to research activity comes in various forms. Most egregiously, misconduct occurs when a scientist presents fabricated research results, or substantially altered experimental data, as the real thing. The case of Eric T. Poehlman provides a prominent example (Kintsch 2005). Formerly a professor at the Vermont College of Medicine, Poehlman admitted in 2005 to falsifying data in fifteen federal grant applications and in numerous published articles dealing with obesity, menopause, and aging. He was considered a leader in his field, and the university had received millions of dollars in federal research support for his work. In 2006 he was sentenced to one year in federal prison, and he is forever barred from receiving any U.S. research funding. The story of his undoing is told in a *New York Times Magazine* article by Jeneen Interlandi (2006). Walter DeNino, a young lab technician working in Poehlman's lab, first reported evidence of fraud to University of Vermont administrators. DeNino was no ordinary lab technician. At the time he was in training for a place on the USA Triathlon team, and he had plans to go on to medical school. When he sensed problems with Poehlman's handling of current research results, DeNino went back to the raw data files that were supposed to form the basis of work already published. It became evident to him that Poehlman had falsified much of the data reported. The university began an investigation that led to a review by the Office of Research Integrity at the Department of Health and Human Services.

The Office of Research Integrity, which oversees research in the biomedical sciences, reports that there are perhaps a dozen cases each year of misconduct of the sort perpetrated by Poehlman. The inspector general of the NSF handles up to five serious cases each year. Those numbers are a minuscule percentage of the thousands of research grants awarded each year in those agencies. The dissemination of fraudulent research results is evidently rare. One reason for this is that any would-be perpetrator of scientific fraud knows that work sufficiently interesting to claim the attention of other researchers will be scrutinized. If it challenges existing assumptions and deals with important issues, others will look closely at the details, and may be moved to conduct experiments on their own to verify the work. Several years ago the authors of a string of papers emanating from Bell Laboratories, the famous industrial research lab, made extraordinary claims of having achieved much sought-after goals in the field of electronic device technology. Researchers in other laboratories therefore pored over the new work in their attempts to understand how such spectacular results were obtained. Suspicion was aroused when they noticed that published data curves for supposedly different systems were identical. When notified of these discrepancies, Bell Labs management appointed an outside committee of experts to investigate. Eventually, after much investigation, the panel found that a single scientist, Jan Hendrick Schön, had acted alone in falsifying data in some twenty-five published papers, most of which had to be withdrawn. It cleared all the other coauthors of the papers of wrongdoing, but noted that some of them may have failed in their responsibility for ensuring the correctness of the reported results. This particular episode probably played an important role in prompting the APS, which represents the nation's forty thousand physicists, to issue a revised and expanded set of ethical guidelines (APS 2002).

Many such stories of research fraud discovery might be told. How much misconduct of this kind exists in science? The truth is that we do not really know. The peer review process employed by reputable journals to ensure that only work of a certain quality is published is not well suited to sniffing out fraud. Rather, it is designed to evaluate work for overall competence and interest for the field. The vast majority of published scientific work is not read by many other scientists, and much of it will never be cited by any other author. Most work adds a little to the store of scientific knowledge but is not sufficiently important or interesting to merit rechecking. Fabrication of results in papers of this kind could easily go undetected altogether. Thus, fraud in the production and publication of scientific papers doubtless exists at some level. The fact is, however, that the public is not much concerned about this kind of scientific misconduct. In spite of occasional high-visibility cases, science enjoys a generally solid reputation for truthfulness and reliability.

There are at least two reasons why this might be so. First, as mentioned above, the detection of error and fraud within science is not regularized and reliable. Scientists themselves cannot give a truly credible account of the frequency of such ethical lapses. In the absence of information to the contrary, fraud does not appear to be a big problem. Second, when significant fraud is uncovered in science, it has nearly always been brought to light through the actions of other scientists or laboratory workers. The public thus has an understanding that science polices itself. Importantly, it is precisely when scientific claims are exceptional—presentation of an entirely unexpected set of observations, inventing a new tool, or challenging well-established theories and models—that the reported work is met with the most skepticism and subjected to the closest scrutiny. (The work from Bell Laboratories is a good example of this process.) Thus, social practices within science go a long way to ensure that fraud and misrepresentation, where it might be most influential, will not long escape detection.[3]

## Misuse of Scientific Knowledge

There are many ways in which scientists can violate ethical codes of conduct by misusing information in their possession. Most of these involve actions that are more or less contained within the boundaries of science. For example, a scientist comes across a terrific new idea in the course of reviewing a grant proposal submitted to a funding agency by a bright young scientist. He promptly instructs a graduate student or postdoctoral research worker to implement the idea in his own laboratory. This very serious ethical violation might escape notice altogether, but if it is uncovered, it will be seen as a problem for the scientific community to deal with. On the other hand, ethical failings that have more direct implications for the public good provide fodder for the public media. As an example, the *Seattle Times* published an article dealing with the leakage of confidential information on drug trials to Wall Street investment firms (Timmerman and Heath 2005). The practice involves payments to medical researchers for information on the progress of ongoing drug trial studies. The article cited as an example the case of Affinitak, an innovative drug developed by Isis Pharmaceuticals for treatment of lung cancer. Unbeknownst to the company, stock analysts had started making calls to physicians who were testing the drug. They received information that led them to believe that the trial was not going well. The news caused Isis stock to spiral downward over a short period of time. Eventually the official results of the trial confirmed that the drug was a failure. Those who had advance warning had of course sold the stock short, and made quick profits of up to 30 percent.

Doctors involved in drug trials sign confidentiality agreements that they will not divulge *any* information about the course of the trial. Why would the doctors in the Affinitak instance be tempted to violate the law? One reason might be that they were paid in the range of three hundred to five hundred dollars per hour to talk with analysts. Even when the calls were not predicated on the doctor divulging confidential information, skillful analysts were able to piece together insights gathered from several calls to

3. Not everyone agrees. See, for example, Fuller 2000.

judge the likely success or failure of a drug trial. The growth of hedge funds, which are not as closely regulated as mutual funds, has made for an anything-goes attitude with respect to gaining an edge in predicting stock price changes. According to the *Seattle Times*, firms now exist whose service is to provide investment analysts with access to doctors. Hedge funds pay such firms up to one million dollars a year for premium access. The doctors who sign up to be available know that the analysts are looking for information that will give them an advantage in judging the near- term movements of key pharmaceutical and biotech stocks. The temptation to violate strict confidentiality in the guise of conveying expert opinion or advice is great.

Ethical failings of this kind are sure to diminish the trust of the public in science. Such self-dealing weakens the moral authority of the biomedical community, but it may also raise questions about the reliability of the results of the studies themselves. Might not the data be "cooked" to produce a desired result? Precautions are taken to run large-scale trials as double-blind studies, in which neither the investigators nor the study patients know whether they are receiving the test drug or a placebo. If the drug produces side effects (for example, a rash) that are evidenced during the study, however, it is possible to infer which patients are which. In any case, the public is not sophisticated about these matters, and news of self-dealing on the part of investigators certainly weakens both their expert and moral authority with the public.

## Conflicts of Interest and Commitment

Scientists can abuse their positions in certain situations by engaging in activities that are in some ways incompatible with their primary responsibilities, or that draw them away from those responsibilities to a detrimental degree. It can happen, for example, when academic researchers consult too often for industrial firms. Excessive time away in consulting detracts from the time and energy available for fulfillment of primary responsibilities. Consulting may also affect the quality and direction of the academic scientist's research activities by directing attention toward problems related to the consulting work and away from more fundamental and creative research. In addition, there has been a steady increase in entrepreneurial activity on the part of faculty researchers. Many have started companies based on patented technology developed in the academic laboratory. The potential for ethical conflicts in such situations is manifold. When the companies are successful the faculty member can become quite wealthy, and in the process new sorts of stresses on professional relationships with colleagues arise.

Conflicts of interest and commitment can also arise in contexts outside academe. One such case involved scientists in the NIH. The NIH provides large amounts of funding for biomedical research in academic institutions and research institutes. In addition, it also operates a huge array of research facilities staffed with life scientists as well as physicians and physician-researchers. Many of them are among the leading figures in their fields. In 2003 and 2004, the *Los Angeles Times* published stories about consulting arrangements enjoyed by certain government scientists employed at the NIH. The articles alleged that the scientists were involved in lucrative consulting agreements with drug companies. Hundreds of NIH scientists took millions of dollars in fees and company stock. In some cases the scientists spoke in public forums or in other ways endorsed particular drugs or treatment plans without revealing their ties to the drug companies that benefited from the recommendations they were making.

The NIH's ethics rules had been relaxed in 1995 in response to an Office of Governmental Ethics audit, which found that the agency's policies were more stringent than those of other executive branch agencies. There was concern at the time that if more opportunities for earning outside income were not available to the top scientists and administrators at NIH, they would leave the agency for employment elsewhere. The relaxed rules removed time and financial limits on outside consulting that did not conflict with the person's official duties. The newspaper stories prompted sweeping investigations by committees of both the House and Senate. At the same time, NIH director Elias A. Zerhouni appointed a blue-ribbon panel to recommend changes in NIH ethics guidelines. The congressional hearings revealed many consulting agreements between NIH scientists and pharmaceutical and biotech companies of which the NIH had no record, a major embarrassment for NIH administration. The congressional findings and the advice of the blue-ribbon panel resulted in the February 2005 issuance of more restrictive guidelines. All NIH employees are barred from participating in paid or unpaid outside activities, not only with pharmaceutical or biotech companies, but also with a long list of other entities. Senior NIH personnel must divest themselves of holdings in excess of fifteen thousand dollars in any one pharmaceutical or biotech company or other companies that might be affected by NIH actions or policies. A final set of provisions was approved and promulgated in August 2005 (National Institutes of Health 2006).

The expectations for scientists employed in corporate laboratories are that they will honor the ethical and moral tenets of science while at the same time serving the demands of their employers. Traditionally, the academic research scientist has been seen as providing the model for disinterested, curiosity-driven study of the natural world. In more recent times, however, the public image of the basic research scientist laboring away in an ivory tower, free from the pressures of the marketplace, has been replaced by something less idealized. Society has increasingly demanded that scientists show how their work connects to the larger society, and what applications might eventually flow from their studies (Sonnert and Holton 2002). Those societal expectations, however, raise the potential for ethical conflicts. The scientist as adviser to government, consultant to industry, inventor of patentable discovery, or even as entrepreneur becomes caught up in the demands of maintaining confidential business relationships and meeting performance goals. These are usually distinct from, and sometimes conflict with, the ethical obligations of the professor who is at once a basic research scientist motivated to engage in curiosity-driven research and a mentor to young scientists striving to establish their own careers.

## THE POWER OF THE PUBLIC

The many surveys of public attitudes show that there is overwhelming support for science and technology as central features of modern society. Yet there is an evident undercurrent in the public sentiment of something like fear mixed with distrust. Science and technology reach into many corners of every person's life; there are bound to be occasions on which conflicts arise between the claims and demands of science and technology and other assertions of what is right and good. We have seen many examples of such contests for authority throughout this book: the competing views of science and religion with respect to human origins; the authority of an epidemiological study evaluating the incidence of ill effects caused by a medication versus individual personal experience; economic arguments against taking a particular precautionary measure related to a pollutant against epidemiological evidence of adverse health effects; and the arguments of some bioethicists and religious conservatives in opposition to stem cell research involving human embryos against the promise of finding treatments and even cures for widespread illnesses.

The vast majority of those in society, though not scientifically trained, are nonetheless obliged to form opinions on the sorts of issues just described if they wish to participate in civic affairs. The public opinion that results is of immense importance to the prosperity and general progress of science and technology. Notice that the issues that come before the public may concern pure science or technology, or some mixture of the two. Whether it is in the best interest of society to permit planting the seeds of genetically modified crops is a question concerning the appropriate application of a particular technology, a downstream by-product of basic research. By contrast, whether stem cell research involving human embryos should be permitted is an issue relating to the governance of research at the stage of basic discovery. The public is not necessarily cognizant of the distinction, and may not care much about it, but it is weighty within the community of science.

The laws that regulate research or its applications, or the funding for particular areas of scientific or technological research, are largely in the hands of government legislative and regulatory authorities, at levels ranging from the federal to state and local levels. We saw in the story of rDNA research that local citizen groups and city councils have the power to limit scientific activity. Yet in terms of their acquaintance with science, elected officials mirror the general public; very few legislators have either a scientific or engineering background. Further, it is relatively rare that legislative staff members have scientific training or background. Under the aegis of the AAAS, a coalition of scientific societies funds about three dozen congressional staff internships to support scientists who wish to participate in public policy, but this represents only a small effort in relation to the need. In the regulatory agencies there is typically an abundance of scientifically and technologically knowledgeable workers. Policy, however, is more often than not controlled by political appointees with no technical credentials.

As science enters into value-laden areas, it becomes increasingly more important for scientists and scientific institutions to engage the broader society on the ethical, legal, social, and economic implications of science and technology. What sorts of values are embraced by science? How do those values find expression in the directions in which science moves, the methods that it employs in working toward its objectives, and the motivations that drive scientific work? The extent to which people trust science—their views about whether science has contributed to the betterment of society as a whole and their optimism about the potential of science to make for a better world—vary with educational achievement, religiosity, and gender (Gaskell et al. 2005).

Communicating information about science is limited by the public's small store of scientific knowledge and understanding of many scientific concepts. A more critical factor, however, is that people generally have only a limited sense of how science is actually practiced, or of how mainstream scientific opinion is formed. While the public undoubtedly lacks a useful understanding of many particular areas of science, such as genetics or atmospheric science, ignorance of the underlying science need not bar acceptance of science's expert authority in those areas. What can and often does constitute a barrier is the absence of an adequate working model in the public imagination of how science actually comes by the knowledge that it claims to have (Yankelovich 2003). Without this understanding, the science of the public imagination is a caricature created largely by media outside science (Ezrahi 2004, chap. 13; Nowotny 2005). The public knows little about real scientists and instead draws its understanding of them from the novels of writers such as Michael Crichton, or from versions of historical figures such as Einstein as presented on TV.

## FRAMING SCIENCE

No one contests that science and technology are important to modern societies. But if science is to play more than a limited instrumental role as a purveyor of information and methods, a role that would seriously restrict its authority and autonomy, it must find ways to engage society more effectively. It must do so in the face of a limited public scientific literacy even as it strives to improve on it. It must reassess the methods that it employs to communicate with the public, especially on controversial matters involving science. The concept of *framing* provides a tool for examining the ways in which science and individual scientists can address various audiences. Erving Goffman, one of the most imaginative of American sociologists, first proposed the use of framing in analyzing social relations and communications (Goffman 1974). He was concerned with finding a general way of thinking about the many ways in which an individual interacts with the social world. Goffman looked at social institutions such as hospitals, schools, or baseball games, and various forms of conventional social activity, such as getting married or going shopping. He saw that people think about institutions or events in terms of mental structures, or frames, which determine how they behave. Frames bear a close relationship to what Lakoff and Johnson refer to as experiential gestalts, or image schema (G. Lakoff and M. Johnson 1980, chap. 15; M. Johnson 1987, chap. 3). It is as though life were a series of little plays or dramas. Frames provide a central organizing idea, a story line, that imparts meaning. For example, the wedding frame has clearly defined roles and staging: the bride, the groom, the father of the bride, the church or outdoor garden outfitted for the wedding, the wedding dress. The participants all have recognized roles and typically fill those roles more or less on cue. Outside the wedding frame the roles have no particular import. Flora is "the bride" on the wedding day; the next day she is back to being just Flora. The frame for an institution also has structure and roles. For example, the frame for a high school contains the corridors with student lockers, the classroom, the gym, and so on. The school librarian, student, and school administrator play the major roles, and there are many supporting roles, such as janitor and cafeteria personnel. In these frames there is a structure and internal logic that connects the physical elements and the roles. Certain types of behavior are deemed appropriate and other types are ruled out. Frames, of course, do not exist in isolation, but are coupled to the environs in various ways. Weddings take place in particular locations, and individual schools have characteristics that distinguish them from one another. But abstracted from the particulars of institutions and events are the structured mental constructs, or experiential gestalts, that convey the general features making up the frames.

We have been speaking of frames for well-established institutions and events, but frames are created continuously as our experiences coalesce around consistent features. Goffman put it this way: "Given their understanding of what it is that is going on, individuals fit their actions to this understanding and ordinarily find that the ongoing world supports this fitting. These organizational premises—sustained both in the mind and in activity—I call the frame of the activity" (1997, 158).

Framing is important in thinking about science's interfaces with society (Nisbet and Mooney 2007a, 2007b). What frames do people employ in responding to key words or phrases such as "biotechnology," "food additive," "chemical research," or "clinical study"? To the extent that the frames consist of impersonal facilities—laboratories, hospitals, heavy equipment at work, and so on—as opposed to individuals talking about their work, or some other more personal style of presentation, science and its applications remain at a remove that precludes establishing communication and trust.

Whereas scientists might suppose that communications with the larger society are most effectively maintained simply by presenting the technical aspects of a scientific finding or methodology in the most user-friendly language possible, the public does not absorb science in this manner. We have already alluded to the limitations posed by the low scientific literacy of the citizenry. This factor, coupled with the challenges that people face in digesting floods of information on all topics, causes people to look for ways to absorb scientific information in terms of existing frames, ones that more often than not have no direct relationship to science. As a simple example, Montana wildlife officials recently reported a significant increase in the number of verified wolf packs and individual wolves in the northwestern part of the state (Associated Press 2007). To many environmentalists this was a bit of good news. The frame they employ in digesting the news is one of wildlife preservation. For officials in the federal Department of the Interior, the frame employed is one of management; the study provided support for a decision to remove the wolf from the endangered species list. To the ranchers of Montana, employing a frame of wolf as predator, the study is an ominous indication that they stand to lose more livestock. The same modest bit of data presented by the state's wildlife officials is incorporated into different frames by different segments of the public, and interpreted in terms of values appropriate to each frame. *Simply put, data and other scientific findings are often not objectively neutral entities; they connect to peoples' worlds in terms of value-laden frames.*

The public's views of science as an endeavor can be thought of in terms of two types of frames. One class includes frames for assimilating the performative aspects of science: a mental construct for how science is actually done; of science as a source of reliable knowledge; of science as a source of solutions to problems that afflict society. In this class there are also frames that encompass other value-laden aspects: science's positive contributions to the cultural life of society; its participation in establishing social values; the trustworthiness of scientists to act in the best interests of society; of scientific information as free of self-dealing or bias. For each of these positive frames, there is the potential for a negative frame; for example, that scientific information is largely biased and serves special interests, or that scientists are largely indifferent to the effects of science on society at large. The general frames that people employ predispose them toward or away from acknowledging expert or moral scientific authority on particular matters that come before them. Someone who believes that scientists generally lack concern for the social consequences of their work is likely to be dubious at the outset about an announced scientific result, and less likely to acknowledge the moral authority of science with respect to those results.

A second set of frames is more specific to the content of scientific reports. As we saw in the example of the wolf study, even when there is no serious question about the validity of the data, the lens through which the data are seen determines their reception in the public arena. When scientists report on the results of stem cell research involving human embryos, on work pointing to global warming, or on data relating exercise programs to cognitive capacities in an aging population, the frames within which such work is evaluated are determined by the interests and background values of those receiving the information. Where new scientific findings present challenges to the prevailing social order, the scientific community should not simply turn over the "facts," as it were, and let society make of them what it will without providing active guidance as to how the new information can be understood in relation to the old. In prior chapters we have seen examples of conflicts that have arisen when new scientific results have threatened entrenched economic interests, as in the case of global warming, or the ethical or religious sensibilities of particular groups, as in the case of stem cell research. An important tactic employed in response has been to weaken the authority of science by attacking key frames that undergird its epistemic and moral authority. One of the most salient of these has to do with the concept of uncertainty.

All scientific knowledge is inherently uncertain in one sense or another. Data have associated with them some level of uncertainty based on how they were obtained. Theories may be incomplete in one respect or another, and thus uncertain to a degree. They might also just be wrong. It may not be feasible at a given time to distinguish between two different interpretations of a certain body of evidence. In communicating about science with the public, it is very important that uncertainty be presented as an inherent element (Friedman, Dunwoody, and Rogers 1999). If science is framed as a source of definitive knowledge, as it so often is in the public mind, an appearance of uncertainty weakens its authority. Scientists are thus tempted to present their results for public consumption without explicitly acknowledging uncertain aspects. They are then left somewhat flat-footed when the public media and special interest groups choose to highlight uncertainties, particularly those arising from disagreements among scientists.

This last strategy has been very popular in the past several years in connection with debates over global warming. The Bush administration and energy corporations such as ExxonMobil have worked diligently and spent a great deal of money to disseminate frames of "scientific uncertainty," "lack of consensus," and "damage to the U.S. economy as a result of taking action" (Robinson 2007; Union of Concerned Scientists 2007). The scientific establishment in this case has made the proper response: acknowledge the uncertainties and talk about what is being done to reduce them. The frame of "worldwide scientific consensus" based on the steady accumulation of new results has increasingly captured the public imagination. It has been made especially prominent by extensive media coverage of the April 2007 release of the report of Working Group II of the IPCC, which deals with the predicted effects of global warming (IPCC 2007c). The uncertainties are clearly presented; for example, the predicted global temperature in the decade 2090–99 is predicted on the basis of one particular model to be 2.8 degrees Celsius higher than in 1990–1999, with a likely range of 1.7 to 4.4 degrees Celsius. This large range of "likely" change reflects a conservative evaluation of all the uncertainties in the modeling and future developments in society. In this example, science is framed as an entity along the lines of Polanyi's "republic of science": the scientific results are presented as a highly vetted scientific opinion, the product of much critical thinking, comparison, and integration of diverse findings, as opposed to the views of individual scientists.

In response to changes in public perceptions, states have taken action to control greenhouse gas emissions (Sunstein 2007, chap. 1). Newly formulated economic arguments have drawn attention to frames of "opportunity" and "innovative technologies," countering the older frame of "huge costs of abatement." The Bush administration's policies are increasingly seen in terms of frames like "protecting special interests" and "political ineptitude." We can thus see that frames are powerful tools for shaping the reception of scientific information in the public arena. In the global warming debate, a remarkable "tipping point" in public opinion has occurred, even in the absence of substantial new scientific evidence bearing on the reality of global warming, its causes, or its predicted effects. Even making allowances for the somewhat mysterious nature of tipping points (Gladwell 2000), the change in the frame employed by much of the public in thinking about the reality of future global climate change is due in large measure to science's well-publicized progress in reducing uncertainties in modeling the future climate. I should emphasize that we are dealing in this instance with science's epistemic authority, and not with issues of what actions, if any, should be taken to mitigate the effects of global warming.

The battle for public acceptance of embryonic stem cell research has similarly been fought in terms of frames, as noted by Nisbet and Mooney

(2007). Those favoring the use of human embryos in nonreproductive cloning work frame their advocacy in such terms as "treatments and cures for diseases" and "competition with other nations for scientific break-throughs." Those opposed talk in terms of frames such as "destruction of human life" and the "amoral hubris of scientists." As this book is being written, the pendulum of public opinion seems to be swinging toward allowing a greater scope of research involving human embryos.

It is important to see that those who initially report scientific findings, whether related to climate change or involving the use of embryonic stem cells, do not generally go out of their way to employ frames evoking value-laden ideas. For example, a scientific paper that reports newly increasing rates of movement of glaciers may entail a relationship to climate change, but the conclusions of the paper are about glaciers. While a motivation of the work might have been to study the effects of climate change on glaciers, that is not the principal thrust of the paper. Scientists reporting on re-search that involves embryonic stem cells are not likely in the published work to attempt a moral defense of the use of such cells. Framing science becomes important in proceeding from the original scientific work to com-munication at a less technical level, one in which the broader social ramifi-cations of the work are brought to the fore. This second step is not typically one that scientists take, although it might occur in a press conference or similar public occasion if the work is of high public interest. James Thom-son of the University of Wisconsin could not have anticipated the intensity of the controversies arising from his initial discoveries involving human embryonic stem cells (discussed in chapter 6), nor wished for the conse-quent notoriety (Kolata 2007). Typically, framing in terms of interest to the public is left to other sectors, many of them outside science itself. Sci-entific institutions such as the American Association for the Advancement of Science, the American Chemical Society, and other professional groups make efforts to frame science favorably in public communications. They support programs that reach out to inform children, parents, and the gen-eral public about science generally and about particular topics, such as health or the environment. But the most important interpreters to the public of the raw materials of scientific discourse are the media. Every day in the newspapers, on the evening news, on Web sites such as Google and Yahoo!, and at many other Internet forums, scientific findings reported in conferences and scientific journals are being presented to the public in forms and contexts that bear few evidences of the details, equivocations, provisos, and possibly even conclusions of the published work. What the

public thus sees of science is a bowdlerized version, on which an ideological bias of one shade or another might be layered (Ezrahi 2004).

The relationship of science to the public is interwoven with its relationships to law, government, religion, and other social sectors. This is true
because the people who serve in special roles in each of these sectors are at
the same time members of the general public. Lawyers, judges, ministers,
churchgoers, members of Congress, mayors, businesspeople, advocates
for causes of all sorts, professors, and convicted felons all have relationships to science, though they may not consciously dwell on them. One set
of frames they employ in thinking about science determines whether they
are disposed to accept the expert authority of science generally: Is science
a reliable source of truth? Is it trustworthy? Commitments to other frames
originating in religion, public culture, and other domains of life determine
each individual's readiness to accept the expert or moral authority of science on any specific issue that brings science into contact, and sometimes
hence into conflict, with those other domains.

In its struggles to gain acceptance and understanding within the public
sphere, science suffers from a systemic disadvantage: very few children, as
part of their early development, encounter ways of thinking that are integral to scientific reasoning or practice. In contrast with exposures to a host
of other influences that shape their understanding of the world, both physical and social, they remain unacquainted with the habits of mind and
characteristic reasoning processes that form the basis of scientific inquiry,
"a process of asking questions, generating and pursuing strategies to investigate those questions by generating data, analyzing and interpreting those
data, drawing conclusions from them, communicating those conclusions,
applying conclusions back to the original question, and perhaps following
up on new questions that arise" (Sandoval 2005). The general features of
such a multistep process, employed in varying degrees and contexts in everyday reasoning, constitute the essence of a naturalistic, rational way of looking at the world. Unfortunately, young learners are exposed to the processes
of orderly observation and causal reasoning relatively late in their intellectual development. Further, inculcating scientific habits of mind seems to be
of lower priority in formal science education than memorized learning of
facts and rules (Alberts 2008, 2009a, 2009b). The result is that people generally are too little imbued with science as a way of thought and as an integral part of culture to generate frames that adequately represent science.

Given these built-in impediments, in order to gain a broader acceptance
and higher level of trust generally, scientists and sciences' spokespersons

must actively seek to increase public acceptance of frames that portray scientists as responsible and engaged members of society, and science as a uniquely valuable source of knowledge. A broader recognition of such frames will not, of course, arise spontaneously; it can come only from increased commitments on the part of individual scientists to engage with the larger society. Only from a base of trust and communication can scientists and science gain a favorable reception for frames that convey epistemic and moral authority on specific issues. While the foregoing discussion may seem to offer little ground for optimism, Ruy Teixeira (2008) has noted encouraging signs in the results of the most recent polls of public opinions about science. Whatever the limitations of their understandings of its workings, people seem overwhelmingly to accord science a great deal of cultural authority on many important issues. The challenge for science is to build on that in establishing a deeper basis for understanding and trust.

# NINE

I have claimed in this book that it is possible to discern two kinds of authority for science. One of these is expert, or epistemic, authority: the capacity to convincingly speak about features of the natural world. The second is moral authority: the license to argue convincingly about how the world should be. To properly understand science's roles in society, we need to distinguish between these two forms of authority, and at the same time recognize their interdependence. Related to the concepts of expert and moral authority is that of autonomy, the capacity for self-determination. For example, the autonomy of the individual scientist in choosing research projects and pursuing promising discoveries is deemed vital to the advance of science. For science as a social institution, autonomy in such matters as credentialing, evaluation of research claims, assignment of credit, recognition of outstanding work, and adjudication of questions relating to fraud and ethical misconduct is at the heart of the scientific ethos.

Our exploration of these themes has shown that it is not difficult to justify a general scientific expert authority. The extent to which science and technology have demonstrated a reliable understanding of the physical world, and the manifold instrumental benefits that have accrued to society, are ample evidence of a vast expertise. The transformative effects of science and technology on the conditions of life in Western society have created an air of authority for science that penetrates deeply into many aspects of daily life. At the same time, our account has revealed that science's epistemic authority has limits. When conflicts with other sources of authority arise, scientifically based rationality often fails to carry the day. Religious beliefs, mistrust of science, rejection of science's value system, and conflicting cultural, political, or economic commitments are among the many factors that limit acceptance of science's claims.

With whatever limitations may attend it, the authority of science generally serves to sustain the expert authority of individual scientists as they communicate their results, both within and outside the boundaries of science as a social sector. We have seen that expert authority is a necessary element in establishing a moral authority for science, but it is insufficient by itself. What is needed in addition is the belief that scientists and the scientific institutions and communities within which they function are committed to improving the welfare of society. This means active participation on the part of scientists in societal affairs, not merely a general assertion that science and technology have led to societal benefits (Yankelovich 2003).

In this final chapter I want to integrate conclusions drawn from foregoing chapters with certain other considerations not yet examined. Our particular concern is with American science. We saw in chapter 3 that the growth and development of science in the United States was very much shaped by the fact that the nation itself was in a nascent stage. There was an emphasis on the practical, an enthusiasm for new ventures. Science was valued more for what it could contribute to economic development and geographical expansion than for its contributions to intellectual life. As science took root, however, with the establishment of research universities and institutions in which basic research was emphasized, scientists followed the lead of European nations in insisting on science for science's sake. Thus, the stage was set for contrasting images of science that have persisted to the present day: science as a bounded community with internally driven goals and standards of both admission and performance, and science as an outwardly attentive community with multiple enduring ties to the rest of society. We begin with a brief return to the historical roots of American science, and move from there to an assessment of how science performs as an element of democratic society.

## THE PRAGMATIC TURN

What would a modern-day Alexis de Tocqueville write about the state of science in the United States? Certainly it would be a vastly different assessment from what the young Frenchman wrote on the basis of his nine-month visit to the United States in 1831. From what were the merest beginnings of scientific study in the early nineteenth century, American science has become an extensive, varied enterprise with enormous global influence. But despite all this success, there are tensions. Surprisingly perhaps, they were foreseen by de Tocqueville from his early vantage point. Americans, he observed, were very much inclined to the practical. One of the sections of *Democracy in America* is tellingly entitled "Why Americans Are More Attracted to Practical Rather than Theoretical Aspects of the Sciences." He argued that there were bound to be conflicts over the aims of scientific study. In part they would arise because of economic and social conditions in the rapidly expanding nation that would place a premium on practical solutions to problems affecting economic growth and social change. He believed also that some of the tensions would arise because of the nature of democratic societies:

> The desire to use knowledge is not the same as the desire to know. I am quite sure that, here and there, some men possess a burning and inexhaustible passion for the truth which is self-supported and a constant source of joy, without ever reaching any final satisfaction.
>
> The future will demonstrate whether such rare, creative passions are born and develop as rapidly in democratic as in aristocratic societies. My own opinion is that I can hardly believe it. (de Tocqueville 1835, 532, 533)

He goes on to make the case that in democratic societies the emphasis will be heavily on the practical and applied aspects of science. We saw in chapter 3 how deeply that sentiment ran in American culture during the nineteenth century as science and technology became important in economic and intellectual life. But as the scientific community grew and became a social entity in its own right, it moved toward a culture grounded in the ideals of disinterested, curiosity-driven research. These shifts conflicted with the expectations of those in industry and government, who viewed science as simply a tool toward economic and social progress. The debate over the proper role and place of science and technology in society continued into the twentieth century. It was clearly reflected in the writings of the pragmatist philosophers, mainly C. S. Peirce, William James, and John Dewey, who created the first truly American brand of philosophy. In very general terms, the pragmatists believed that knowledge is instrumental;

its function is to integrate, predict, and control our interactions with the experienced world. Science represented a particularly powerful methodology for acquiring useful knowledge. John Dewey, the most widely influential of the pragmatist philosophers, was concerned with how to place science in society in light of pragmatist ideas (Dewey 1939, 457). Throughout his writings Dewey views science, along with the humanities and other areas of intellectual activity, as having both a private character and a broader social function. The conflict between science's internal values and goals and the demands that society makes on it has continued to play a role in defining the compact between science and society.

Vannevar Bush's *Science: The Endless Frontier* was powerfully influential in setting the tone of the compact following World War II. It was premised on the notion that science and technology would perform basic research that would meet the needs of defense-related projects, and that it would provide a reliable pool of trained scientists in key scientific disciplines. In return, research funds became freely available for research chosen by scientists themselves, in accord with their professional judgments. Bush's goal had been to secure a measure of autonomy for curiosity-driven research—in other words, freedom from the strictures that came with funding for targeted ends. He largely succeeded in that effort. As Donald Stokes has written, "It was widely accepted in the postwar years that basic science can serve as a pacemaker of technological progress only if it is insulated from thought of practical use" (1997, 27). The Russian launch of Sputnik in 1957 only served to reinforce the notion that technological developments of the future depend on fundamental breakthroughs in the present. Despite entanglements with national security issues, the relationship with the federal government worked well for quite a few years—long enough to establish the validity of peer review of research proposals and merit-based awards of research resources in the minds of nonscientists.

For a time science was just where it wanted to be: there was adequate funding for research (though scientists themselves almost never believe that to be the case) and wide latitude in choices of research projects. At the same time, something important was missing. Broadly speaking, there was little or no public voice in science policy or governance. The public became involved with science only when particular matters, such as the safety of rDNA research, or effects on public health from environmental contamination by pesticides, rose to the fore. Small wonder, then, that when the political climate changed dramatically in the 1970s, and science came under pressure to justify its requests for research resources in terms of its practical contributions to society, it was not successful in mustering widespread public support.

The period from about 1955 to 1970 provides insights into what can happen when basic research scientists—mostly academics—and their parent institutions are relatively free to construct the world of science. Plentiful governmental funding, a laissez-faire attitude on the part of funding agencies regarding research choices, a good job market for those trained as research scientists, institutional encouragements, and many other factors together made for a community with remarkably few specific ties to the larger society. It was at just this time that the novelist and scientist C. P. Snow (1959) drew attention in his famous Rede Lecture to what he saw as a great cultural divide in society between science and technology on the one hand and arts and letters on the other. There was plenty of political activity on the part of scientists acting as advisers in Washington, and among institutions competing for shares of the federal largesse for science. Infighting over allocations to disciplinary areas was at times intense, and charges of elitism were common (Kevles 1995, chap. 24). Public participation in science-related legislation or in policy discussions with scientific or technological content, however, was limited, except in matters related to science education in the schools. Most academic research scientists, who typically made many scientific presentations to peer groups in other institutions and at scientific meetings of all sizes, had few occasions to present their work to lay audiences, nor did most of them seek such opportunities. In terms of the model of science outlined by Polanyi in his famous 1962 paper, quoted in chapter 4, this all seemed pretty much as it should be. Polanyi argued that scientists must be free to make choices of research projects, and that they should be sheltered from distractions:

> The only justification for the pursuit of research in universities lies in the fact that the universities provide an intimate communion for the formation of scientific opinion, free from corrupting intrusions and distractions. For though scientific discoveries eventually diffuse into all people's thinking, the general public cannot participate in the general milieu in which discoveries are made. Discovery comes only to a mind immersed in its pursuit. For such work the scientist needs a secluded

place among like-minded colleagues who keenly share his aims and sharply control his performances. The soil of academic science must be exterritorial in order to secure its rule by scientific opinion. (74)

Polanyi's model of basic science is inherently elitist. There is nothing surprising in this; the pathway to a career in scientific research is arduous, and relatively few people are equipped in terms of attitude or aptitude to embark on it. At the heart of the arguments made by Polanyi and others for scientific autonomy and elitism is that they are justified by the nature of the scientific enterprise itself. The degree of autonomy that Polanyi called for has never been fully realized by science in democratic societies such as the United States and Britain, but aspects of his vision are shared by most scientists. To illustrate, in 1975 a committee of the AAAS issued a report entitled *Scientific Freedom and Responsibility*, written on behalf of the committee by a distinguished biochemist, John T. Edsall. The committee was formed in 1970, at a time when there was much controversy over abridgements of scientists' freedom of expression on controversial social issues. The text of the report reflects a romantic image of the scientist as a dedicated searcher for truths about nature:

> The central value for the most creative scientists is a passionate dedication to the advancement of knowledge, and to the solution of the baffling puzzles with which nature confronts us. The desire, for instance, to discover the principles underlying quantum phenomena at the atomic and subatomic level, or to understand the chemical nature of the gene and its replication, has preoccupied some of the greatest scientific minds of this century. Such workers often live with their problems, day and night, for months or even years on end; the desire to solve a particularly baffling problem may keep them preoccupied, to an extent that most other people can scarcely conceive. A deep sense of the beauty of science, and its aesthetic appeal, commonly accompanies this urge to solve difficult and fundamental problems. Indeed, the aesthetic appeal of science is very strong for many other, less creative researchers; but this appreciation of the beauty of science requires in general an arduous preliminary training that makes it unattainable for most nonscientists. (AAAS 1975)

Although science has changed greatly in the intervening three decades, many scientists today would probably not find this account outlandish.

Notice, however, that it has nothing to say of the scientist's connections with the social world outside science. For us a question of interest is, How could such an insular scientific community exercise authority?

## THE CHANGING STRUCTURE OF SCIENCE

We observed in chapter 4 that individual scientists generally see themselves as belonging to tightly knit communities with characteristics that support claims for authority. The trouble is that while their perceptions and the social structure on which they depend have significance for those immersed in the world of science, their meaning is obscure for nonscientists, who are generally unfamiliar with that world (Yankelovich 2003). Furthermore, they do not answer to criticisms that science has evolved in a culture that privileges some at the expense of others (Keller 1985; Keller and Longino 2006; Burke and Mattis 2007). Thus we might question whether the cognitive and moral virtues that Robert Merton identified (discussed in chapters 3 and 4) actually speak to contemporary concerns about the nature of science. Beyond this, however, it is fair to ask whether or to what extent they continue to reflect and actively help to shape the social structure of the scientific community. Enormous changes have occurred in the social structure of science in recent decades (Guston 2000). This is not the place to examine them in any detail, but simply mentioning a few provides hints of the ways in which science's boundaries with society are currently shifting, with important implications for its capacity to exercise authority.

First, an increasing number of scientists occupying faculty positions and performing research in other settings are foreign-born. Most, but no means all, have received their advanced scientific training in the United States. According to the NSF's *Science and Engineering Indicators, 2008*, 41 percent of the doctorates in science and engineering positions in the United States in 2005 were foreign-born, as were more than 20 percent of the science and engineering faculty in U.S. universities (National Science Foundation 2008). Notably, the percentage of foreign-born scientists has been increasing steadily. The results raise many questions: Just how truly universal is the culture of science? How does the culture of science couple with larger societal cultures? Will it make a difference to the authority of American science that a large fraction of all basic research scientists and engineers are the products of cultures other than American?

Second, women and minorities have long been underrepresented in American science. Despite a variety of programmatic efforts to address the problem, the percentage of minority scientists remains far below their representation in the population at large (AAAS 2007). In recent years the percentages of women in many scientific disciplines has increased to the point where they are no longer underrepresented among those graduating with both undergraduate and graduate degrees. Yet women do not advance in recognition in proportion to their numbers. For example, election to the NAS is one of the highest honors that an American scientist can receive. The percentage of women among the nearly two thousand members of the Academy stands at about 6.5 percent. Feminist scholars have pointed to the many ways in which the culture of science disadvantages women. At a conference on the role of women and minorities in sciences, Bruce Alberts, president of the NAS, said: "We all recognize that science is, and must be, an elitist enterprise. It needs our very best minds. How do we bring more people into this enterprise? It is very important that we do so for many reasons. One often talks about the unfairness of not giving every-body a chance to contribute. But an even bigger issue in this country, as it becomes more and more diverse, is that a science establishment run pri-marily by white males runs the danger of alienating our nation and our people from science" (NAS 2000). Alberts has identified an important ten-sion: between the culture of science as an elitist institution with stringent entrance requirements and a male-oriented tradition of competitiveness, dominance, and control on the one hand, and a societal demand for inclu-siveness on the other. Failure of the science establishment to resolve this tension significantly diminishes its potential for exercising moral author-ity on many matters.

Finally, the traditional model of the academic scientist isolated in the ivory tower, working to uncover the mysteries of nature in relative isolation from the distractions of the marketplace, no longer holds in many disci-plines (Bok 2003). Today academic institutions conduct research efforts in collaboration with industrial partners, and ownership rights in intel-lectual property are central elements in contractual arrangements. The range of industries involved is wide, encompassing biotech, pharmaceuti-cal, agricultural, electronics, and defense-related companies. In other cases faculty members form start-up companies to develop commercial products from inventions made in the university research laboratory, which are then patented by the university and licensed to the faculty member. Or the in-vention may be licensed to an existing company, which pays royalties on income if the invention proves to be a commercial success. The numerous stories of faculty members who have become multimillionaires through such activities serve to fuel a perception that Merton's criteria of communal solidarity and disinterestedness have faded in importance (Shapin 2008, chap. 7). The perception that basic science is increasingly devoted to the search for commercially successful products, especially when the scientists directing the research are direct beneficiaries, threatens both science's epistemic authority and moral authority. Scientists who have commercial interests in particular areas of science and technology can hardly be counted on for objectivity in resolving issues that arise in those areas, or that in any way might affect them.

### PUBLIC PARTICIPATION IN SCIENCE

Surely one of the important factors in determining the authority roles that science can play in American society is the degree to which the larger society perceives science as being *in and of* society. If science is seen as exclusive and narrowly professional in its composition and outlook, it will receive only a grudging acquiescence in matters relating to its expert authority, and little or no response to its efforts to exercise moral authority. One of the great challenges for science is thus to be more open to the influences of the larger society. But how is this to be accomplished? Some believe that broadening the mandate for basic science will go a long way toward meeting legitimate societal demands for an increased focus on society's needs (Branscomb, Holton, and Sonnert 2002, 397–433; Sonnert and Holton 1999; Stokes 1997). But even if changes along those lines are implemented, how are they to be brought about? During the latter half of the twentieth century and since, the scientific community has urged that the best choices in the long run will be made within science, based on the standards of the community and through a process of arriving at a consensus based on values intrinsic to science.

For many, simply broadening the aims of basic research, or somehow elevating applied science in the hierarchy of scientific value, is not responsive to a larger challenge of making science more democratic (Guston 2000, 2004; Kitcher 2001; M. Brown, 2004). Greater access to science by women and minorities, monitoring of conflicts of interest, establishing appropriate ethical standards, enhancing educational outreach, testing the objectivity of scientific advice on matters relating to the public good, and many more such concerns engage the larger society in important ways. Some argue for more public participation in the affairs of science, including decisions as to which areas of science should be supported, and perhaps even which research projects should be funded. Science is being asked to contribute not only to the advancement of knowledge, but also, more directly than ever before, to the solution of important societal problems, such as creating a sustainable economy and environment; enhancing the use of natural resources through exploration, conservation, and recycling; enhancing security at the personal and national level; and fostering a stronger civic culture (AAAS 2008a, 2008b). More active roles for science in these areas will require commitments from individual scientists to work outside traditional boundaries.

But this brings us face-to-face with a predicament. As we saw in chapter 8, the scientific literacy of the public is low. Even among the better-educated public a woeful fraction of nonscientists have a basic understanding of scientific concepts. How then can the public become productively involved in making decisions that relate to allocating resources for support of scientific research, or shaping policies that govern research practices? The public can and does become engaged—for example, through political activism, participation in public hearings, and membership on public panels (M. Brown, 2004). But such activities do not in themselves build an understanding of the scientific issues that lie at the heart of decision making about science.

The obstacles to substantial public participation are formidable. They consist in more than just lack of a specific scientific background; scientists themselves often participate in science planning activities that involve matters far removed from their own areas of expertise. More important than specific content knowledge is the tacit knowledge involved in understanding how science works: making observations and interpreting them; formulating hypotheses and testing them; evaluating the prospects for new discovery in what is known (Polanyi 1958, pt. 2; 1966). Tacit knowledge plays a role in planning what to do next. It is a product of actually performing scientific research, and therefore is not available to a nonscientist with a limited knowledge of scientific concepts and no experience performing as a scientist.

Despite these difficulties, a greater role for the public in setting goals and limits for science must be found. If the public does not acquire a more informed understanding of how priorities for science might be determined, there will be no public spokespersons for science as it faces challenges to its autonomy and authority. Philip Kitcher attempts in his book *Science, Truth, and Democracy* to formulate social arrangements for decision mak- ing that would set the agenda for science with the informed participation of all segments of society, producing what he terms "well-ordered science" (2001, chap. 10). Kitcher envisions a role for "deliberators," who would come together to evaluate the significances of various proposed courses of action and make recommendations for commitments of resources to sci- entific projects. He says little, however, about who these deliberators might be, even though the feasibility of his scheme turns on successfully execut- ing this first step. The idea seems to be that they are deliberating to gen- erate a list of socially desirable outcomes that they would like science to provide. Quite surprisingly, he does not seem to expect that the delibera- tors will have much understanding of science. The next stage, getting from the lists of desirable outcomes to the allocations of resources to particular investigators and particular projects, would, he suggests, require the inputs of experts.

Kitcher's proposals raise many questions. What reasons do we have for thinking that scientifically naive deliberators, bringing to the proceedings only their particular set of goals for society, will be able to agree on goals that can be successfully mapped onto specific areas of scientific practice? Something closer to models in current practice would seem to afford more promise. First, people with experience in setting scientific priorities and evaluating the merits of competing interests within science, in the light of larger societal interests and goals, are uniquely able to exercise authority with government and the public on issues relating to science. These delib- erators could have scientific backgrounds, but that would not be necessary. If they are to be at all effective in conversing with scientists in setting pri- orities, however, they must be well-informed on positions within the science establishment on the issues that come up for discussion. Second, they must have taken the time and effort to learn about how science actually works: the social structure of science, the factors that motivate individual scien- tists, the elements of good scientific practice, how work in science gets evaluated, and so on. Third, they must be as free as possible from conflicts and biases that would compromise their judgments on contested issues, and be prepared to declare all potential conflicts of interest as they arise. Fourth, they must be capable spokespersons, willing to communicate with the public on what they have learned, and prepared to provide recommen- dations—that is, prepared to exercise a moral authority that has as its basis their willingness to work in the public interest.

Where are deliberators to be found, and how might they be appointed and put to work? One avenue is to work through agencies of government, ranging from the most local to state and perhaps even federal levels. Impressively large numbers of people are willing to volunteer for boards and committees devoted to civic purposes, nearly always without any remuneration. Given the importance of science in society, appropriately constituted committees devoted to science policy and oversight should be able to attract able people. There could, for example, be a "science board" at the level of state government to advise the governor and state legislators on matters relating to science policy and funding within each state. Many state-level boards already exist, though they are primarily focused on specific topics. For example, the state of North Carolina has a Board of Science and Technology that advises on the role of science and technology in the state's economic growth and development. The Michigan Environmental Science Board, appointed by the governor, is charged with providing advice to the governor and state departments, as requested by the governor, on matters affecting the protection and management of Michigan's environment and natural resources. The California Air Resources Board, appointed by the governor with consent of the state senate, is a rule-making body dealing with energy and environmental concerns. All of these boards have membership places for representatives of the public as well as for knowledgeable scientists and engineers. They are, however, narrowly focused. What is needed are science boards with broad charters and membership drawn eclectically from among the public. The governor would be responsible for appointing the board. She would need to exercise care that the board did not consist simply of a group of scientists, retired or otherwise, or of people pushing narrow agendas. Advice on membership would need to be solicited from the science and business communities, public interest groups, civic organizations, and others. Those with experience on such state science boards could in time constitute a source of informed and capable membership for panels and boards at the federal level. They could also operate at more local levels when appropriate, perhaps through appointments of subcommittees.

As examples of its activities, a state science board might be asked to (a) examine the state's annual budget with the aim of evaluating components with significant science content; (b) evaluate a proposal to fund a program dealing with an environmental issue; (c) examine proposed legislation regarding conflicts of interest for scientists employed by the state; (d) evaluate a proposal to establish a center for post-genomic research at one of the state universities; or (e) comment on proposed legislation regulating the use of water from aquifers to irrigate farmland. Do the programs and proposals appear to be based on good science? What will be their likely impact in the public domain? Are they well structured to accomplish their scientific objectives? Do they complement or compete with existing initiatives?

These sketchy ideas merely hint at one general direction that might be taken to broaden public participation in the affairs of science. They do not, however, address the goal of engaging a large proportion of the lay public with issues relating to science. I am unconvinced that there is a way to accomplish this, given the low state of civic literacy with respect to science, and generally low level of interest in scientific matters. Communication with the general public through the agency of such boards could, however, become one more avenue for imparting knowledge and values about science. It would provide a means of framing science more realistically with respect to specific issues of public concern, and be more convincing for having come from largely nonscientist-appointed boards.

Answering the questions that such publicly constituted boards might pose, or otherwise serving as a resource for them, might be seen by many working scientists and engineers as an unwelcome drain on their professional lives. Science and engineering organizations might see governmentally appointed entities as competition with their own efforts to communicate science to the public, advocate for their constituents, or have a voice in making public policy. But at stake here is more than communication or advocacy, important as that is. The most important function of such participatory boards would be to help set agendas for science, and to frame science more favorably in the public domain. The gain for science would be a greater public awareness of what science contributes to society, with a concomitantly greater readiness to accept scientific authority.

## THE LIMITS OF SCIENTIFIC AUTHORITY

It will be valuable at this point to consider what we can say about the limits of scientific authority. To begin we might ask how the public, that undifferentiated multitude that forms society, evaluates science. We know that the public has a rather low scientific literacy. People do not know much about science and often harbor beliefs that contradict those held consensually within science. Nevertheless, they don't seem to take those contradictions very much to heart. They express generally positive feelings toward science; 86 percent of those in the Virginia Commonwealth University survey mentioned in chapter 8 judged that science and technology have helped to make the world better.

Children's directly embodied experiences in early childhood create deep-seated conceptual frameworks that guide thought (Bloom and Weisberg 2007). Some of these conceptual frameworks conflict with scientific knowledge, making certain principles difficult to learn. In this way, resistance to some scientific concepts may be a strong human propensity. People are seldom forced to confront such misconceptions, which thus have little consequence for their views about science. Certain misconceptions that persist into adulthood, however, may stand in the way of accepting particular scientific claims, particularly about human origins and aspects of human behavior.

When people are made aware of scientific evidence that conflicts with beliefs that are closely tied to cultural values, they generally resist changing their outlooks. A prime example is the low rate of public acceptance of evolution (Miller, Pardo, and Niwa 2007). The claim that humans are evolved from earlier forms of life is resisted not only by biblical literalists, but also by many Catholics and mainstream Protestants in the United States. In a cross-national survey, the percentage of respondents who believe the tenets of evolutionary theory was far lower in the United States than in any of the nine European countries surveyed. The apparent cultural bias likely stems at least in part from ignorance of the science; extensive survey work indicates that adults in the United States have a poor awareness of genetic concepts. Large numbers of people pass through their secondary school education without significant exposure to the principles of genetics, or for that matter any science whatever. In summary, there is little scientific knowledge in society. Furthermore, scientific claims often seem to be at variance with cultural predilections and values, particularly those involving religious beliefs.

## CRITIQUES OF SCIENTIFIC AUTHORITY

When what science has to say contravenes traditional ways of thinking, or threatens to upset settled interests, there is resistance, and science is forced to defend itself. Attacks that originate in competing social camps often take the tack of denigrating scientific authority. Critics generally rely on several central criticisms, some that apply to the particular case at hand, and some to science generally.

*First, positions or actions that science advocates conflict with those based on the dictates of traditional authorities.* This kind of challenge to scientific expert or moral authority is perhaps the most resistive to a rational resolution. It implicates appeals to an authority other than science—to core values imparted through family life during childhood, and reinforced in the daily social life of a community. Important public policy conflicts surrounding such topics as abortion, stem cell research, sex education, and the teaching of evolution in schools create battlegrounds on which scientific perspectives clash with conservative religious and political views. There may be small victories in these conflicts, such as the Dover School Board decision described in chapter 6, but longer-term gains are realized mainly through slow cultural changes that occur as one generation succeeds another. One should not be too optimistic about some kinds of changed outlooks. For example, opposition to teaching the concepts of evolution is continually reinforced by its having been made an element of conservative national politics, so it retains a salience that it might not otherwise have (Miller, Pardo, and Niwa 2007).

*Second, there is a lack of consensus in the scientific community.* We have seen repeatedly that scientific authority relies on perception of a general consensus within the relevant scientific community on the validity of key observations and the conclusions drawn about them. Unanimity is not required, but rather that a very large majority of those working in the field are in agreement. Opponents attempt to weaken the authority of a position embodied in a scientific consensus by disproportionately highlighting the views of dissenters. This tactic is effective, in part, because nonscientists do not have the tools needed to make considered judgments about the relative worth of contested opinions offered by scientists.

*Third, insufficient information is available to permit an informed decision.* This argument is commonly advanced in resistance to scientific evidence bearing on complex environmental and health issues. For example, fossil fuel industry interests, in conjunction with the Bush administration, have insisted for many years that we lack sufficient information to be confident about the origins of global temperature change, or even whether one was in the offing. Thus we should do nothing, or at most carry out further research. This tactic can provide an effective argument; there is always more to learn, and scientists tend to couch their findings in language that recognizes the existence of uncertainties: We should wait until we are more confident of the facts. We need more "sound science" (Mooney 2005, chap. 6).

*Finally, science lacks authority because of ethical failures.* The face of science has been changing dramatically over the past few decades. Many of those changes have placed scientists in positions where there are more temptations than in the past to transgress moral and ethical canons. Any evidence of wrongdoing on the part of practicing scientists serves to weaken the authority status of science. The defining characteristics of the scientific community as delineated by Robert Merton lack the force they once had, in part because science has become so much more fragmented and diverse in terms of its couplings with outside interests. The medical researcher working with patients in a drug trial, the academic chemist collaborating with a pharmaceutical firm in designing a new drug, the physicist studying the properties of potential new superconducting substances, the team of social scientists and health professionals evaluating the effects of nutrition on learning in children: all have particular and diverse challenges in maintaining ethical behavior and avoiding conflicts of interest. There exists no single set of rules, limitations, and constraints on practice that can apply to scientists and scientific organizations generally.

## CENSORING SCIENCE

Though science might have the potential for contributing to the resolution of an issue of public importance, it cannot do so if it lacks the means for making itself heard. For example, certain scientific perspectives can be excluded from legislative hearings by those who control the access and do not wish to give voice to the consensus scientific position. Senator James Inhofe largely kept mainstream scientists out of hearings on global warming of the Committee on Environment and Public Works for the many years he served as chair, in favor of a small cadre of dissenters with limited credentials. Political appointees in regulatory agencies have systematically deleted from scientific reports language that refers to facts or assessments that they do not wish to see presented.

Scientific information may be excluded from primary and secondary school classrooms by school boards or state committees that screen texts for suitability. Among the better-known text screening committees are the textbook review panels in the state of Texas, which evaluate instructional materials submitted for potential use in the state. The panels are charged with evaluating materials to determine coverage of essential skills and to identify errors. Texas law states that the board can reject books based only on "factual errors." Eventually the panels report through the commissioner of education to the Texas State Board of Education. The materials are rated as conforming, nonconforming, or rejected. A report detailing any factual errors to be corrected in instructional materials prior to delivery to school districts is also presented.

Disagreements over textbook contents frequently become a flash point between conservatives and liberals, particularly with respect to the topics of evolution, sex education, global warming, and environmental science. The review panels are under pressure from various interested parties to force publishers to alter content. For example, conservative Christian groups have insisted that books dealing with human reproductive health emphasize abstinence and avoid discussion of contraceptive methods. Their influence has resulted in substantial changes in health science texts (Elliott 2004). According to the *New York Times*, seven of the fifteen members of the Texas State Board subscribe to the notion of intelligent design (Beil 2008). Don McLeroy, a dentist who has chaired the State Board, rejects evolution and believes that Earth's appearance is a recent geological event. In October 2008, the State Board of Education moved to appoint three new members to a six-member panel that is to review new science curriculum standards for the state. The three were characterized by Alan Leshner, CEO of the AAAS, as antievolution activists (Leshner 2008). They are: Stephen Meyer, a senior fellow and vice president of the Discovery Institute; Ralph Seelke, a science professor at the University of Wisconsin-Superior; and Charles Garner, a Baylor chemistry professor. Meyers and Seelke are the authors of *Explore Evolution*, which has been widely criticized by science organizations for creating misimpressions about evolution and promoting the institute's intelligent design agenda. All three have signed a statement promoted by the institute expressing skepticism about evolution (Leshner 2008). Do the actions of such boards constitute censorship of science, or are they legitimate processes for ensuring that textbooks are accurate and balanced in treating sensitive topics? There are no simple litmus tests that such panels can apply; as a result, politics, religious beliefs, and other cultural forces easily override a scientific perspective.

We saw in chapter 5 that through questionable application of the criteria for admissibility of scientific evidence established by the Supreme Court in the *Daubert-Joiner-Kumho* decisions, the court system may at times exclude scientific testimony that could be powerfully probative. The *Daubert* court established a two-pronged test for admissibility of scientific evidence: it must be both relevant and reliable. Among the characteristics to be looked for in evaluating reliability are that the hypotheses motivating the study were properly tested; that the work had been peer-reviewed and published in scientific journals of acceptable quality; that the techniques employed were subject to standards of evaluation and were generally recognized in the relevant scientific community; and that the known or potential rates of error were accounted for.

It is up to the trial judge to determine whether these criteria have been satisfied. But to apply these criteria the judges must have a reasonable grasp of science. Without it they have no way of determining whether scientific evidence based on new advances in science might be useful, nor of assessing the value of an epidemiological study that might or might not accord with other lines of evidence. It seems clear that there is great respect for science in the courts; the challenge is how to use it well. If science is to be more effective in legal proceedings, evaluations for admission of scientific evidence must accord more faithfully with the knowledge and practices of science, and be more adaptable to advances in science. At this point it is not clear how these goals might be accomplished, but it seems desirable that the courts have access to more unbiased scientific advice than they do at present.

## ADVANCING SCIENTIFIC AUTHORITY

From all the foregoing it is evident that scientists and scientific organizations make a multitude of important contributions to broad social discourses dealing with a wide range of contentious topics. Whether the issue be one involving climate change, the effects of coal-burning power plants on human health, military preparedness, the regulation of food and drug distribution, sex education, the teaching of evolution, regulation of animal care in research facilities, or the admissibility of certain evidence in the courtroom, the scientific perspective is likely to be at its heart. Nevertheless, it is but one perspective, vying with others for attention and acceptance. What can scientists and science organizations do to enhance science's authority? Clearly it is very important to pay close attention to the ways in which any scientific issue is framed, and, in the same vein, how science itself is framed. Nonscientists will more openly accept scientific assertions, particularly when they contrast with settled opinions, if they have a better grasp of how science functions as a social community. This does not require that they be able to understand in any detail the scientific principles and practices that produce a particular result. Rather, it has to do with knowing how scientific opinion is formed, with having confidence in the methods of science generally. How does any individual scientist stand in relationship to his or her discipline, and that discipline to the whole of science? How did the claims being put forth come to be accepted as a best approximation to the truth, not only by the scientist presenting them, but by a substantial fraction of the scientists concerned with this area of science? In short, science must continually work at more effectively explaining itself to society—not just in terms of how it comes by particular results, but also in terms of how its way of proceeding leads to reliable knowledge. Yet in some situations these considerations will seem to be largely beside the point, because they do not adequately acknowledge the powerful grip of early life experiences, which I have remarked on elsewhere in the book. Consider the stories of two men whose early lives were quite similar in many ways, but who took different paths later in life.

Pastor Rick Warren, a Christian evangelical minister, is one of the most widely read and followed religious leaders in the United States. Born in 1954 in San Jose, California, and raised partly in Texas, he is the son of a Baptist minister. He received a Bachelor of Arts degree from California Baptist University in Riverside, a Master of Divinity degree from Southwestern Baptist Theological Seminary in Fort Worth, and a Doctor of Ministry degree from Fuller Theological Seminary in Pasadena, California. He leads the twenty-thousand-member Saddleback Church in Lake Forest, California, and has written the best-selling book *The Purpose-Driven Life.* In his writings and preaching he encourages people and churches to look for and be motivated by God's purpose for them, as opposed to acting out of fear, a slavish devotion to tradition, or various material considerations. In recent years he has increasingly encouraged evangelicals to get involved in addressing large-scale societal problems such as poverty and the AIDS epidemic. Despite his immense popularity, Warren has been criticized by many evangelical Christians because of his advocacy for social action. When he invited Barack Obama to speak at a 2006 global AIDS conference held at his Saddleback Church, it was taken by many religious conservatives as evidence of his liberal political tendencies.

The April 9, 2007, issue of *Newsweek* contained an account of a dialogue moderated by editor Jon Meacham between Rick Warren and Sam Harris, an avowed atheist (Harris 2006), on the subject "Is God Real?" The discussion wanders over many questions and topics, and quite predictably comes to no satisfying conclusions or agreements. One of the questions asked and Warren's answer, however, are significant for our concerns here:

> JON MEACHAM: "Do you believe Creation happened in the way Genesis describes it?"
>
> WARREN: "If you're asking me do I believe in evolution, the answer is no, I don't. I believe that God, at a moment, created man. I do believe Genesis is literal, but I also know metaphorical terms are used. Did God come down and blow in man's nose? If you believe in God, you don't have a problem accepting miracles. So if God wants to do it that way, it's fine with me."

Stephen Godfrey, born in 1960, was also raised in a fundamentalist Christian family, in Quebec, Canada. His parents possessed a religious faith in a "Young Earth," six thousand years in age, in which Noah's flood laid down the fossils. As a youngster Godfrey developed a keen interest in biology, particularly in skeletal and fossil remains. He relates that when he was told by his first-grade teacher that apes were the ancestors of humans, "I remember having this visceral reaction . . . and saying, 'No, that can't be.'" (Couzin 2008, 1035). He lived at home while attending Bishop's University in Sherbrooke, Quebec, majoring in biology. Godfrey encountered many bits of evidence that seemed difficult to explain in terms of his religious convictions, but managed to put these to the back of his mind. When he worked as part of a field expedition during his first summer in graduate school, however, the challenges to his religious presuppositions, posed by his own observations, began to shake his confidence in the Young Earth creationist story. Eventually he found himself unable to sustain a belief in creationism, and broke away from it, to the consternation of his family. Today Godfrey is a paleontologist and curator at the Calvert Marine Museum in Solomons, Maryland. As a creationist-turned-evolutionist, he struggles to maintain the ties to his family and to religious traditions that remain important to him and his wife.

Rick Warren and Stephen Godfrey represent contrasting responses to the challenge of reconciling deeply held religious beliefs with the epistemic claims of science. For Warren, the challenge was never really taken up. His path through higher education was clearly focused largely on theological matters, and it is unlikely that he was ever encouraged to give serious attention to scientific evidence for an ancient earth or for evolution. Nor is it likely that he ever had much exposure to scientific rationality as a mode of thought. Thus, it is relatively easy for Warren to say that he does not believe in evolution. It is simply a name for something others believe that contradicts his deeply held religious beliefs, and is thus not worthy of serious attention. This man, with exceptional talents and an exemplary devotion to good works, has simply made the decision not to engage with scientific claims that appear to conflict with his religious beliefs. Stephen Godfrey, on the other hand, having given way to his inherent curiosity about the natural world, was brought to a place where his warranted beliefs about the nature of the world simply did not square with those depending on religious authority. As with many others, he had to make a painful choice.

By contrasting the experiences of these two men I do not mean to imply that the conflict between the traditional authority of religion and the epistemic authority of science is the only, or even a primary, occasion of conflicts between scientific and other sources of authority. Indeed, the prior chapters of the book provide many examples drawn from other social domains. It does serve as a salient example, however, of how the ideas and values with which we are imbued early in life form a developmental foundation, firmly implanting not only beliefs, but also habits of thought, that inhibit our capacities to absorb ideas and evidence that have the potential to change our outlooks. As I noted in chapter 8, science must overcome its relatively weak role in most people's intellectual formation. It has the difficult challenge of engaging with people on what is typically unfamiliar material brought to bear on questions that they do not fully understand, presenting ideas with the potential for challenging cherished or at least comfortably held beliefs.

It is thus not surprising that in its attempts to exercise expert authority in society at large, scientists frequently encounter cultural impediments to full and unbiased assessments. To add to the difficulties, people are often put off by the manner in which they are confronted with new ideas. For many, the scientist is stereotyped as an "objective" knower, dispassionate and analytical, comfortable in discussions premised on skepticism and challenge. Many people find this stereotype threatening; for them, to understand a phenomenon is to experience it in some way, not simply to accept someone else's word for how things are. We can see evidences for such attitudes in studies of the responses of college students to various learning situations. The studies reveal that habits of mind and personality characteristics determine peoples' capacities to absorb new ideas. For example, Blythe McVicker Clinchy and her colleagues have written of their experiences in teaching women at Wellesley College (Belenky et al. 1986). They found it useful to distinguish "separate" and "connected" modes of knowing on the part of their students. Separate knowers employ strategies typical of scientists; to use a term coined by the influential teacher of writing Peter Elbow (1973), they play a "doubting game," searching for flaws in positions, looking for alternative explanations, and offering competing models. Connected knowers play what Elbow terms "the believing game," attempting to make sense of narratives that may at first seem not to make sense, by empathically extending their belief systems. As students, both these kinds of knowers need to make changes in their ways of interpreting external events and ideas in order to achieve greater objectivity toward new ideas, and the same applies to people generally. But the traditional modes of pedagogy that dominate teaching methods, and that extend as well to the efforts of science to connect with nonscientists, do not facilitate this process. Furthermore, not all students are ready to engage in such a process. Those that Clinchy and her colleagues term "subjectivists" are unwilling to make the effort of reconciling their beliefs with countervailing views presented to them.

The distinguished psychologist Jerome Bruner argues that academic disciplines are dominated by what he calls a "paradigmatic" approach that relies on logical argument and empirical testing for its force. But for Bruner, "Neither the empiricist's tested knowledge nor the rationalist's self-evident truths describe the ground on which ordinary people go about making sense of their experiences" (1996, 130). People come to the point of giving assent to a proposition when they have participated in a process of constructing a narrative, a possible story that accounts for how things are. "We devote an enormous amount of pedagogical effort to teaching the methods of science and rational thought: what is involved in verification, what constitutes contradiction, how to convert mere utterances into testable propositions, and on down the list," writes Bruner. "For these are the 'methods' for creating a 'reality according to science.' Yet we live most of our lives in a world constructed according to the rules and devices of narrative" (149). Clinchy writes, "The emphasis is on making meaning . . . rather than evaluating finished ideas. Impregnable arguments and definitive experiments are irrelevant to this enterprise; personal feelings are not" (Clinchy 2007–8, 27).

My purpose here is not to present a thorough brief for a revised pedagogical approach that will somehow improve the effectiveness of science's communications with those outside science. It is rather to suggest that a turn toward more personal presentations of science, including narratives that relate stories of scientists at all stages of their scientific development, will make science more approachable. More important still is to actively help nonscientists make connections with what is being presented by recognizing that their personal histories and feelings play a role in governing their responses. Inevitably, communicating science becomes more akin to a dialogue than the typical unidirectional mode of dispensing scientific information employed by those attempting to exercise expert authority. A more personal approach that relies on narrative as well as logical argument establishes closer connections with nonscientists, and in so doing enhances science's moral authority.

### FINAL WORDS

I recently attended a production of Arthur Miller's play *The Crucible*. This classic drama is built on the historical facts of the 1692 witchcraft trials in Salem, Massachusetts, which resulted in the conviction and hanging of nineteen men and women. In 1957 the Massachusetts General Court passed a resolution that declared the proceedings to have been the result of popular, hysterical fear of the devil. The episode, so awful to our modern sensibilities, became in Miller's hands the basis of an allegorical tale. It offered a parallel to the political turmoil of his own time, the anticommunist hysteria that fed the House Un-American Activities Committee and Senator Joseph McCarthy's notorious hearings.

One of the characters in the drama is the Reverend John Hale, a minister called to Salem to ascertain whether there is witchcraft. Hale is a young intellectual with a deep belief in the Bible. He is something of a specialist in witchcraft as a result of much reading and pondering on the matter. When he first enters the stage, he is carrying books dealing with witchcraft. Hale's breezy confidence in his abilities to discern witchcraft are sorely tested as more and more people in Salem are caught up in a hysteria that began with the strange behavior of a few young girls. Hale is convinced that the devil is afoot in Salem. At one point, when people are being locked up in droves, he talks with John Proctor, who believes that the adults are being victimized by the lies of a little girl seeking vengeance. Proctor's wife has just been taken away to be locked up awaiting trial. Hale, distraught at the turn of events, exclaims: "Proctor, I cannot think God be provoked so grandly by such a petty cause. The jail are packed—our greatest judges sit in Salem now—and hangin's promised. Man, we must look to cause proportionate. Were there murder done, perhaps, and never brought to light? Abomination? Some secret blasphemy that stinks to Heaven? Think on cause, man, and let you help me to discover it. For there's your way, believe it, there is your only way, when such confusion strikes upon the world" (A. Miller 1954, 74).

Poor Hale! He and the Puritans of Salem, having placed all their faith in literal interpretations of a book, were helpless to defend themselves against wholesale surrender to the irrational. Hale knew that there must be explanations for everything that happens, but he simply had no idea of how to find them. This character in Miller's drama brought home to me how powerfully the modes of inquiry ushered in by the Enlightenment changed human society. The rise of science brought with it new ways of asking questions. The scientific project is to seek causal explanations for what is observed in nature, to test those explanations with experiments, to strive continuously for improved understanding. At its heart is a disciplined skepticism: take nothing for granted. Science, more than any other social force, created modernity, in which the public expression of radical ideas became possible.

Quite aside from its instrumental value in improving the physical conditions of life, science has become a core element of Western culture more generally. Scientists, like anyone else, may hold ideas that are simply wrong. Individual scientists have at times advanced faulty theories regarding race, social status, and gender that served those in power. Scientists have at times been complicit in supporting repressive political regimes. But these shortcomings were typically short-lived, and were brought to an end by actions of the scientific community itself. They must be measured against the enormous positive changes that science has wrought. Scientific discoveries have made it possible to hold progressive ideas about human nature that were previously unthinkable. In American society today, the conversations about such topics as religious beliefs, ageing, homosexuality and gay marriage, the origins of criminal behavior, the deleterious effects of persistent organic environmental pollutants, childhood education, global warming—the list is virtually endless—are all informed in one way or another by perspectives that are the products of scientific investigation.

Science does not provide answers to every important question. It does, however, contribute to the discussions, and it is entirely appropriate that it should do so. Science is indeed an authoritative voice in American society. How could it be otherwise, given that society itself is so thoroughly suffused with the consequences of scientific research? But science and technology can and should do more. Scientists and engineers have been instrumental in creating contemporary American society. They have done so by informing us about how the world is. It is vital to the future of humanity that they more fully apply their understandings and skills to helping us decide how the world should be; to forming a just, liberal, and livable society.

# REFERENCES

BOOKS AND ARTICLES

AAAS. 1975. *Scientific freedom and responsibility.* Washington, D.C.: American Association for the Advancement of Science. http://archives.aaas.org/docs/1975-SFR.pdf.
———. 2007. Gender bias legislation opens discussion on women in science. *Science and Technology in Congress,* November. http://www.aaas.org/spp/cstc/stc/Archive/stc07/07_11_stcnewsletter.html.
———. 2008a. AAAS dialogue on science, ethics, and religion. http://www.aaas.org/spp/dser.
———. 2008b. Forum on science and technology policy. http://www.aaas.org/spp/rd/forum.htm.
———. 2008c. Global climate change resources. http://www.aaas.org/news/press_room/climate_change.
ABC News. 2006. Catholic school teacher fired for having in vitro: School officials said in vitro fertilization contradicted Church doctrine. *Good Morning America,* May 11. http://www.abcnews.go.com/GMA/story?id=1949131.
Alberts, Bruce. 2008. Considering science education. *Science* 319:1589.
———. 2009a. Making a science of education. *Science* 323:15.
———. 2009b. Redefining science education, *Science* 323:437.
All Catholic Church Ecumenical Councils. http://www.piar.hu/councils/~index.htm.
Allchin, Douglas. Galileo trial: 1616 documents. http://my.pclink.com/~allchin/1814/retrial/1616docs.htm.
Althaus, F. 1991. U.S. religious groups vary in patterns of method use, but not in overall contraceptive prevalence. *Family Planning Perspectives* 23 (6): 288–90.
Altimore, M. 1982. The social construction of a scientific controversy: Comments on press coverage of the recombinant DNA debate. *Science, Technology, and Human Values* 7:24–31.
American Association for the Advancement of Science. *See* AAAS.
American Bar Association, Section of Individual Rights and Responsibilities. 2002. http://www.abanet.org/leadership/recommendations02/117b.pdf.
American Medical Association. 2003. Cloning and stem cell research. http://www.ama-assn.org/ama/pub/category/13630.html.
American Physical Society. *See* APS.
Angell, Marcia. 1994. Do breast implants cause systemic disease? *New England Journal of Medicine* 330:1748–49.
———. 1996. *Science on trial.* New York: Norton.
Annas, Julia. 2006. Moral knowledge as practical knowledge. In *The philosophy of expertise,* ed. Evan Selinger and Robert P. Crease. New York: Columbia University Press.
APS. 2002. APS guidelines for professional conduct. http://www.aps.org/policy/statements/02_2.cfm.

———. 2003. Nobel laureates and industry leaders petition president to save U.S. science and technology. http://www.aps.org/about/pressreleases/petition.cfm.

Associated Press. 2005. Court rules on Baby 81. *New York Times*, February 14.

———. 2007. Officials document increase in wolf population. *Billings Gazette*, February 23. http://www.billingsgazette.net/articles/2007/02/23/news/state/43-wolves.txt.

Augustine. 1998. *Confessions*. Trans. Henry Chadwick. New York: Oxford University Press.

Bacon, Francis. 1605. *The advancement of learning, novum organum, and new Atlantis*. Chicago: Encyclopaedia Britannica, 1952.

———. 1620. *The new organon*. New York: Washington Square Press, 1963.

Bala, G., K. Caldera, A. Mirin, M. Wickett, and C. Delire. 2005. Multicentury changes to the global climate and carbon cycle: Results from a coupled climate and carbon cycle model. *Journal of Climate* 18:4531–44.

Barnes, Barry. 1985. *About science*. Oxford: Oxford University Press.

Barnes, Barry, and David Bloor. 1982. Relativism, rationalism, and the sociology of knowledge. In *Rationality and relativism*, ed. Martin Hollis and Steven Lukes. Cambridge, Mass.: MIT Press.

Barnes, David W. 2001. Too many probabilities: Statistical evidence of tort causation. *Law and Contemporary Problems* 64:191–212.

Barrett, James P. 1999. The high cost of distorted economic modeling. *The Journal of Commerce*, February 7. http://www.epinet.org/printer.cfm?id=567&content_type=1&nice_name=webfeatures_view.

Barrett, James P., and J. Andrew Hoerner. 2002. *Clean energy and jobs: A comprehensive approach to climate change and energy policy*. Washington, D.C.: Economic Policy Institute. http://www.clt.astate.edu/crbrown/GlobalWarming8.pdf.

Barton, Joe. 2001. House hearings. http://frwebgate.access.gpo.gov/cgibin/useftp.cgi?IPaddress=162.140.64.21&filename=71503.wais&directory=/diskc/wais/data/107_house_hearings.

Bauer, Linda. 2008. Campaign for seats on Kansas Board of Education runs quiet. *Kansas City Star*, November 1.

Behe, Michael. 1996. *Darwin's black box: The biochemical challenge to evolution*. New York: Free Press.

Beil, Laura. 2008. Opponents of evolution adopting a new strategy. *New York Times*, June 4. http://www.nytimes.com/2008/06/04/us/04evolution.html.

Belenky, Mary Field, Blythe McVicker Clinchy, Nancy Rule Goldberger, and Jill Mattuck Tarule. 1986. *Women's ways of knowing*. 10th anniv. ed. New York: Basic, 1997.

Ben-David, Joseph. 1971. *The scientists' role in society*. Englewood Cliffs, N.J.: Prentice Hall.

Bender, Thomas. 1984. The erosion of public culture: Cities, discourses, and professional disciplines. In *The authority of experts: Studies in history and theory*, ed. Thomas Haskell. Bloomington: Indiana University Press.

Benedick, Richard Elliot. 1991. *Ozone diplomacy: New directions in safeguarding the planet*. Enlarged ed. Cambridge, Mass.: Harvard University Press, 1998.

Berg, Paul, David Baltimore, Herbert W. Boyer, Stanley N. Cohen, Ronald W. Davis, David S. Hogness, Daniel Nathans, et al. 1974. Potential biohazards of recombinant DNA molecules. *Science* 185:303.

Berg, Paul, David Baltimore, Sydney Brenner, Richard O. Roblin III, and Maxine

Singer. 1975. Asilomar conference on recombinant DNA molecules. *Science* 188:991–94.

Berger, Margaret. 2005. What has a decade of *Daubert* wrought? *American Journal of Public Health* 95:S59–65.

Bhattacharjee, Yudhijit. 2005a. Kansas gears up for another battle over teaching evolution. *Science* 308:627.

———. 2005b. Groups wield copyright power to delay Kansas standards. *Science* 310:754.

Biagioli, Mario. 1993. *Galileo: Courtier.* Chicago: University of Chicago Press.

———. 2006. *Galileo's instruments of credit: Telescopes, images, secrecy.* Chicago: University of Chicago Press.

Bieber, Frederick B. 2004. Science and technology of forensic DNA profiling: Current use and future directions. In *DNA and the criminal justice system,* ed. David Lazer. Cambridge, Mass.: MIT Press.

Bird, Kai, and Martin J. Sherwin. 2005. *American Prometheus: The triumph and tragedy of J. Robert Oppenheimer.* New York: Knopf.

Black, Bert, Francisco J. Ayala, and Carol Saffran-Brinks. 1994. Science and the law in the wake of *Daubert*: A new search for scientific knowledge. *Texas Law Review* 72:715–802.

Blackburn, Elizabeth. 2004. Bioethics and the political distortion of biomedical science. *New England Journal of Medicine* 350 (14): 1379–80.

Blackmore, Susan. 1999. *The meme machine.* Pbk. ed. Oxford: Oxford University Press, 2000.

Bloom, David, and David Canning. 2004. Global economic change: Dimensions and economic significance. National Bureau of Economic Research Working Paper 10817. http://www.kc.frb.org/PUBLICAT/SYMPOS/2004/pdf/BloomCanni ng2004.pdf#search='David%20Bloom%20and%20David%20Canning'.

Bloom, Paul, and Deena Skolnick Weisberg. 2007. Childhood origins of adult resistance to science. *Science* 316:996–97.

Bloor, David. 1984. A sociological theory of objectivity. In *Objectivity and cultural divergence,* ed. S. C. Brown. Cambridge: Cambridge University Press.

Bok, Derek. 2003. *Universities in the marketplace: The commercialization of higher education.* Pbk. ed. Princeton, N.J.: Princeton University Press, 2004.

Brady, Catherine. 2007. *Elizabeth Blackburn and the story of telomeres: Deciphering the ends of DNA.* Cambridge, Mass.: MIT Press.

Branscomb, Lewis W., Gerald Holton, and Gerhard Sonnert. 2002. Science for society. In *AAAS Yearbook,* ed. A. H. Teich, S. D. Nelson, and S. J. Lita. Washington, D.C.: AAAS.

Brodie, Richard. 1995. *Virus of the mind: The new science of the meme.* Seattle: Integral.

Bronowski, Jacob. 1956. *Science and human values.* Rev. ed. New York: Harper and Row, 1965.

Brook, Edward J. 2005. Tiny bubbles tell all. *Science* 310:1285–87.

Brown, Mark B. 2004. The political philosophy of science policy. *Minerva* 42:77–95.

Brown, Mark B., and David H. Guston. 2009. Science, democracy, and the right to research. *Science and Engineering Ethics.* http://www.springerlink.com/con tent/765u3t574r224h3h.

Brown, Theodore L. 2003. *Making truth: Metaphor in science.* Champaign: University of Illinois Press.

Bruner, Jerome S. 1996. *The culture of education.* Cambridge, Mass.: Harvard University Press.

Bucchi, Massimiano. 1998. *Science and the media: Alternative routes in scientific communication.* New York: Routledge.

Burke, Ronald J., and Mary C. Mattis, eds. 2007. *Women and minorities in science, technology, engineering, and mathematics: Upping the numbers.* Northampton, Mass.: Edward Elgar.

Bush, Vannevar. 1945. *Science: The endless frontier.* 40th anniv. ed. Washington, D.C.: National Science Foundation, 1990.

Butler, Declan. 2006. Top five science blogs. *Nature* 442:9.

California Assembly. 2006. Assembly bill no. 32. http://www.leginfo.ca.gov/pub/05-06/bill/asm/ab_0001-0050/ab_32_bill_20060927_chaptered.pdf.

Carlson, Elof A. 2006. *Times of triumph, times of doubt.* Cold Spring Harbor, N.Y.: Cold Spring Harbor Press.

Carmen, Ira H. 1985. *Cloning and the constitution: An inquiry into governmental policy-making and genetic experimentation.* Madison: University of Wisconsin Press.

———. 2004. *Politics in the laboratory: The constitution of human genomics.* Madison: University of Wisconsin Press.

Carson, Rachel. 1962. *Silent spring.* New York: Houghton Mifflin.

Cassara, Ernst. 1988. *The enlightenment in America.* 2nd ed. Lanham, Md.: University Press of America.

Cassidy, David C. 2005. *J. Robert Oppenheimer and the American century.* New York: Pi Press.

Catholics for Choice. 2006. *Complying with the law? How Catholic hospitals respond to state laws mandating the provision of emergency contraception to sexual assault victims.* Washington, D.C.: Catholics for Choice. http://www.catholicsforchoice.org/topics/healthcare/documents/2006complyingwiththelaw.pdf.

———. 2008. *Truth and consequences: A look behind the Vatican's ban on contraception.* Washington, D.C.: Catholics for Choice. http://www.catholicsforchoice.org/topics/reform/documents/TruthConsequencesFINAL.pdf.

Caudill, David S., and Lewis H. LaRue. 2006. *No magic wand: The idealization of science in law.* Lanham, Md.: Rowman and Littlefield.

CBS News. 2004. Poll: Creationism trumps evolution: Most Americans do not believe human beings evolved. November 22. http://www.cbsnews.com/stories/2004/11/22/opinion/polls/main657083.shtml.

Chakraborty, R., and K. K. Kidd. 1991. The utility of DNA typing in forensic work. *Science* 254:1735–39.

Chapin, John B. 1880. Experts and expert testimony. *Albany Law Journal* 22:365–76.

Christie, Maureen. 2000. *The ozone layer.* Cambridge: Cambridge University Press.

Clinchy, Blythe McVicker. 2007–8. Beyond subjectivism. *Tradition and Discovery* 34 (1): 15–31.

CNN Politics. 2008. Obama may reverse Bush policies on stem cells, drilling, abortion. http://edition.cnn.com/2008/POLITICS/11/11/obama.executive.orders/index.html.

Coady, C. A. J. 1992. *Testimony.* New York: Oxford University Press.

Cole, Simon A. 2001. *Suspect identities: A history of fingerprinting and criminal identification.* Cambridge, Mass.: Harvard University Press.

Coleman, Howard, and Eric Swenson. 1994. *DNA in the courtroom: A trial watcher's guide.* Seattle: Genelex.

Collins, Francis S. 2006. *The language of God: A scientist presents evidence for belief.* New York: Free Press.

Copeland, Libby. 2006. Faith-based initiative. *Washington Post,* June 7. http://www .washingtonpost.com/wp-dyn/content/article/2006/06/06/AR2006 060601616_pf.html.

Couzin, Jennifer. 2008. Crossing the divide. *Science* 319:1034–36.

Cowan, Chad A., Irina Klimanskaya, Jill McMahon, Jocelyn Atienza, Jeannine Witmyer, Jacob P. Zucker, Shunping Wang, et al. 2004. Derivation of embryonic stem-cell lines from human blastocysts. *New England Journal of Medicine* 350 (13): 1353–56.

Cranor, Carl. 2005. Scientific inferences in the laboratory and the law. *American Journal of Public Health* 95 (S1): S121–28.

———. 2006. *Toxic torts: Science, law, and the possibility of justice.* New York: Cambridge University Press.

Daniel, George. 1971. *Science in American society.* New York: Knopf.

Darwin, Charles. 1871. *The descent of man.* New York: Modern Library, 1936.

Davies, Kevin. 2001. *Cracking the genome: Inside the race to unlock human DNA.* New York: Free Press.

Davis, Edward B. 2005. Science and religious fundamentalism in the 1920s. *American Scientist* 93:253–60.

Davis, Percival, and Dean H. Kenyon. 1989. *Of pandas and people: The central question of biological origins.* 2nd ed. Richardson, Tex.: Foundation for Thought and Ethics, 1993.

Dawkins, Richard. 1976. *The selfish gene.* 2nd ed. Oxford: Oxford University Press, 1989.

———. 2006. *The God delusion.* New York: Houghton Mifflin.

Debré, Patrice. 1994. *Louis Pasteur.* Baltimore: Johns Hopkins University Press.

Dembski, William. 1999. *Intelligent design: The bridge between science and theology.* Downers Grove, Ill.: Intervarsity.

———. 2002. *No free lunch: Why specified complexity cannot be purchased without intelligence.* Lanham, Md.: Rowman and Littlefield.

Dennett, Daniel C. 2006. *Breaking the spell: Religion as a natural phenomenon.* New York: Viking.

Descartes, René. 1637. Discourse on method. In *Man and the universe: The philosophers of science,* ed. Saxe Commins and Robert N. Linscott. New York: Random House, 1947.

Dessler, Andrew E., and Edward A. Parson. 2006. *The science and politics of global climate change: A guide to the debate.* New York: Cambridge University Press.

de Tocqueville, Alexis. 1835. *Democracy in America.* Trans. Gerald E. Bevan. London: Penguin, 2003.

Devlin, B., Neil Risch, and Kathryn Roeder. 1993. Statistical evaluation of DNA fingerprinting: A critique of the NRC's report. *Science* 259:748–50.

Dewey, John. 1939. *Intelligence in the modern world: John Dewey's philosophy.* Ed. Joseph Ratner. New York: Random House.

Dewolf, David K., John G. West, Casey Luskin, and Jonathan Witt. 2006. *Traipsing into evolution: Intelligent design and the* Kitzmiller v. Dover *decision.* Seattle: Discovery Institute.

Dickson, A. C., L. Meyer, and R. A. Torrey, eds. 1910–15. *The fundamentals: A testimony.* 12 vols. Chicago: Testimony.

Dixon, Lloyd, and Brian Gill. 2001. *Changes in the standards for admitting expert evidence in federal civil cases since the* Daubert *decision*. Santa Monica, Calif.: RAND Institute for Civil Justice. http://www.rand.org/pubs/monograph_reports/2005/MR1439.pdf.

Donige, David, and Michelle Quibell. 2007. *Back from the brink: How NDRC helped to save the ozone layer*. New York: National Resources Defense Council. http://www.nrdc.org/globalwarming/ozone/ozone.pdf.

Duke Law. 2003. Dr. Leon Kass presents Siegel memorial lecture. http://www.law.duke.edu/features/news_kass.html.

Durkheim, Émile. 1912. *Elementary forms of religious life*. Trans. Karen Fields. New York: Free Press, 1995.

Elbow, Peter. 1973. The doubting game and the believing game. In *Writing without teachers*. Oxford: Oxford University Press.

Elliott, Janet. 2004. Critics say texts ignore information on contraception. *Houston Chronicle*, June 29. http://www.chron.com/CDA/archives/archive.mpl?id=2004_3776696.

Elsberg, Wesley R. 2005. Apparent end of Dover School Board's reign of error. *Panda's Thumb*, November 9. http://pandasthumb.org/archives/2005/11/apparent_end_of.html.

Ezrahi, Yaron. 2004. Science and the political imagination in contemporary democracies. In *States of knowledge: The co-production of science and social order*, ed. Sheila Jasanoff. London: Routledge.

Federal Register. 2002. Office of Management and Budget guidelines for ensuring and maximizing the quality, objectivity, utility, and integrity of information disseminated by federal agencies. http://www.whitehouse.gov/omb/fedreg/reproducible2.pdf.

Fehring, Richard, and Andrea Matovina Schmidt. 2001. Trends in contraceptive use among Catholics in the United States: 1988–1995. *Linacre Quarterly*, May, 170–85.

Feyerabend, Paul. 1975. *Against method*. 3rd ed. London: Verso, 1993.

———. 1978. *Science in a free society*. London: New Left.

Finocchiaro, Maurice A. 2005. *Retrying Galileo*. Berkeley and Los Angeles: University of California Press.

Fleck, Ludwik. 1986. *Cognition and fact: Materials on Ludwik Fleck*. Ed. R. S. Cohen and T. Schnelle. Dordrecht, Netherlands: Reidel.

Fliri, Anton F., William T. Loging, Peter F. Thadeio, and Robert A. Volkman. 2005. Analysis of drug-induced effect patterns to link structure and side effects of medicines. *Nature Chemical Biology* 1:389–97.

Foster, Kenneth R., David E. Bernstein, and Peter W. Huber, eds. 1993. *Phantom risk: Scientific inference and the law*. Cambridge, Mass.: MIT Press.

Foster, Kenneth R., and Peter W. Huber. 1999. *Judging science: Scientific knowledge and the federal courts*. Cambridge, Mass.: MIT Press.

Frederickson, Donald S. 2001. *The recombinant DNA controversy: A memoir: Science, politics, and the public interest, 1974–1981*. Washington, D.C.: American Society for Microbiology Press.

Friedman, Lee M. 1910. Expert testimony: Its abuse and reformation. *Yale Law Journal* 19:247–57.

Friedman, Sharon M., Sharon Dunwoody, and Carol L. Rogers, eds. 1999. *Communicating uncertainty*. Mahwah, N.J.: Lawrence Erlbaum.

Friedman, Thomas L. 2008. *Hot, flat, and crowded.* New York: Farrar, Straus and Giroux.

Fuller, Steve. 2000. Commentary on Polanyi's *Republic of science. Minerva* 38 (1): 26–32.

Furstenberg, Frank F., Jr. 1994. History and current state of divorce in the United States. *Children and Divorce* 4 (1): 29–43.

Galilei, Galileo. 1615. Letter to the Grand Duchess Christina. http://www.galilean-library.org/manuscript.php?postid=43841.

Gaskell, George, Edna Einsiedel, William Hallman, Susanna Hornig Priest, Jonathan Jackson, and Johannus Olsthoorn. 2005. Social values and the governance of science. *Science* 310:1908–9.

Gatowski, S. I., Shirley A. Dobbin, James T. Richardson, Gerald P. Ginsburg, Mara L. Merlino, and Veronica Dahir. 2001. Asking the gatekeepers: A national survey of judges on judging expert evidence in a post-*Daubert* world. *Law and Human Behavior* 25:433–58.

Geison, Gerald L. 1995. *The private science of Louis Pasteur.* Princeton, N.J.: Princeton University Press.

Gewirth, Alan. 1978. *Reason and morality.* Chicago: University of Chicago Press.

Giere, Ronald. 1992. The cognitive construction of scientific knowledge. *Social Studies of Science* 22 (1): 95–107.

Gieryn, Thomas F. 1999. *Cultural boundaries of science: Credibility on the line.* Chicago: University of Chicago Press.

Gingerich, Owen. 2006. *God's universe.* Cambridge, Mass.: Harvard University Press.

Gish, Duane T. 1979. *Evolution? The fossils say no!* San Diego: Creation-Life.

Gladwell, Malcolm. 2000. *The tipping point: How little things can make a big difference.* Boston: Little, Brown.

Goffman, Erving. 1974. *Frame analysis: An essay on the organization of experience.* New York: HarperCollins.

———. 1997. *The Goffman reader.* Ed. Charles Lemert and Ann Branaman. Oxford: Blackwell.

Golan, Tal. 2004. *Laws of men and laws of nature: The history of scientific expert testimony in England and America.* Cambridge, Mass.: Harvard University Press.

Goldberg, Steven. 1994. *Culture clash.* New York: New York University Press.

Goldman, Alvin I. 1999. *Knowledge in a social world.* Oxford: Clarendon.

———. 2006. Experts: Which ones should you trust? In *The philosophy of expertise,* ed. Evan Selinger and Robert P. Crease. New York: Columbia University Press.

Golinski, Jan. 1998. *Making natural knowledge: Constructivism and the history of science.* Cambridge: Cambridge University Press.

Goodchild, Peter. 2004. *Edward Teller: The real Dr. Strangelove.* Cambridge, Mass.: Harvard University Press.

Goodstein, Laurie. 2006. Evangelical leaders join global warming initiative. *New York Times,* February 8. http://www.nytimes.com/2006/02/08/national/08warm.html.

Gopnik, Adam. 2006. Rewriting nature. *New Yorker,* October 23, 52.

Gough, Michael. 1993. Dioxin: Perceptions, estimates, and measures. In *Phantom Risk,* ed. Kenneth R. Foster, David E. Bernstein, and Peter W Huber. Pbk. ed. Cambridge, Mass.: MIT Press, 1999.

Gould, Stephen Jay. 1997. Nonoverlapping magisteria. *Natural History* 106 (March): 16–22.

Government Accountability Project. 2006. GAP announces new watchdog, "Climate Science Watch." http://www.whistleblower.org/content/press_detail .cfm?press_id=405&keyword=.

Grant, Edward. 2006. *Science and religion, 400 B.C. to A.D. 1550: From Aristotle to Copernicus.* Baltimore: Johns Hopkins University Press.

Green, Michael D. 1992. Expert witnesses and sufficiency of evidence in toxic substances litigation: The legacy of Agent Orange and Bendectin litigation. *Northwestern University Law Review* 86 (3): 643–99.

———. 1996. *Bendectin and birth defects.* Philadelphia: University of Pennsylvania Press.

Greenland, Sander, and James M. Robins. 2000. Epidemiology, justice, and the probability of causation. *Jurimetrics Journal* 40:321–40.

Griffin, David Ray. 2002. Naturalism: Scientific and religious. *Zygon* 37 (2): 361–80.

Grobstein, Clifford. 1979. *A double image of the double helix.* San Francisco: Freeman.

Groscup, Jennifer L., Steven D. Penrod, Christina A. Studebaker, Matthew T. Huss, and Kevin M. O'Neil. 2002. The effects of *Daubert* on the admissibility of expert testimony in state and federal criminal cases. *Psychology, Public Policy, and Law* 8:339–72.

Guston, David H. 2000. *Between politics and science.* Cambridge: Cambridge University Press.

———. 2004. Forget politicizing science: Let's democratize science! *Issues in Science and Technology* (Fall): 25–28. http://www.cspo.org/ourlibrary/articles/ DemocratizeScience.htm.

Haack, Susan. 2003. *Defending science—within reason: Between scientism and cynicism.* New York: Prometheus.

———. 2005. Trial and error: The Supreme Court's philosophy of science. *American Journal of Public Health* 95 (S1): S66–73.

Haber, Lyn, and Ralph Norman Haber. 2003. Error rates for human latent fingerprint examiners. In *Advances in Automatic Fingerprint Identification*, ed. N. K. Ratha and R. Bolle. New York: Springer-Verlag.

Habermas, Jürgen. 2008. *Between naturalism and religion.* New York: Polity.

Hacking, Ian. 1983. *Representing and intervening: Introductory topics in the philosophy of natural science.* Cambridge: Cambridge University Press.

Hale, George Ellery. 1922. A national focus of science and research. *Scribner's Magazine*, November 22, 515–31.

Hansen, James. 2005. Is there still time to avoid "dangerous anthropogenic interference" with global climate? Presented December 6 at the annual meeting of the American Geophysical Union, San Francisco, Calif. http://www .columbia.edu/~jeh1/keeling_talk_and_slides.pdf.

———. 2007. Why we can't wait. *The Nation*, May 7. http://www.thenation.com /doc/20070507/hansen.

Hansen, James, Larissa Nazarenko, Reto Ruedy, Makiko Sato, Josh Willis, Anthony Del Genio, Dorothy Koch, et al. 2005. Earth's energy imbalance: Confirmation and implications. *Science* 308:1431–34.

Hardwig, John 2006. Epistemic dependence. In *The philosophy of expertise*, ed. Evan Selinger and Robert P. Crease. New York: Columbia University Press.

Harmon, Rockne. 1990. How has DNA evidence fared? Beauty is in the eye of the beholder. *Expert Evidence Report* 1.

Harris, Sam. 2006. *Letter to a Christian nation.* Pbk. ed. New York: Vintage, 2008.

Harvey, William. 1628. On the motion of the heart and blood in animals. http://www.fordham.edu/halsall/mod/1628harvey-blood.html.

Hilgartner, Stephen. 2000. *Science on stage: Expert advice as public drama.* Stanford, Calif.: Stanford University Press.

Hobbes, Thomas. 1651. *Leviathan.* New York: Oxford University Press, 2008.

Hoffman, David I., Gail L. Zellman, C. Christine Fair, Jacob F. Mayer, Joyce G. Zeitz, William E. Gibbons, and Thomas G. Turner Jr. 2003. Cryopreserved embryos in the United States and their availability for research. *Fertility and Sterility* 79 (5): 1063–69.

Hofstadter, Richard. 1963. *Anti-intellectualism in American life.* New York: Vintage.

Holden, Constance. 2006a. Darwin's place on campus is secure—but not supreme. *Science* 311:769–71.

———. 2006b. States, foundations lead the way after Bush vetoes stem cell bill. *Science* 313:420–21.

Holden, Constance, and Gretchen Vogel. 2008. Cell biology: A seismic shift for stem cell research. *Science* 319:560–63.

Holton, Gerald. 2000. Coupling science and the public interest. Lewis Branscomb Lecture, Harvard University, March 16.

Hooper, Laural J., Joe S. Cecil, and Thomas E. Willgang. 2001. Assessing causation in breast implant litigation: The role of scientific panels. *Law and Contemporary Problems* 64:139–89.

Horrigan, John B. 2006. *The Internet as a resource for new and information about science: The convenience of getting scientific material on the Web opens doors to better attitudes and understanding of science.* Washington, D.C.: Pew Internet and American Life Project. http://www.pewinternet.org/pdfs/PIP_Exploratorium_Science.pdf.

Howard Hughes Medical Institute. 2004. New embryonic stem cell lines to be made available to researchers. http://www.hhmi.org/news/melton4.html.

Huber, Peter W. 1993. *Galileo's revenge: Junk science in the courtroom.* New York: Basic.

Hughes, Thomas P. 2004. *The human-built world: How to think about technology and culture.* Pbk. ed. Chicago: University of Chicago Press, 2005.

Hume, David. 1748. *An inquiry concerning human understanding.* Ed. Tom L. Beauchamp. Oxford: Oxford University Press, 1999.

Huxley, Thomas H. 1859. Letter to Charles Darwin, November 23. http://www.ucmp.berkeley.edu/history/thuxley.html.

Intergovernmental Panel on Climate Change. *See* IPCC.

Interlandi, Jeneen. 2006. Unwelcome discovery. *New York Times Magazine,* October 22. http://www.nytimes.com/2006/10/22/magazine/22sciencefraud.html.

International Association for Identification. 1980. Resolution VII amended. *Identification News* 3 (August).

Inwood, Stephen. 2002. *The man who knew too much: The strange and inventive life of Robert Hooke, 1635–1703.* London: Macmillan.

IPCC. 2007a. *Fourth assessment report: Climate change 2007 assessment report.* Geneva: Intergovernmental Panel on Climate Change. http://www.ipcc.ch/pdf/assessment-report/ar4/syr/ar4_syr.pdf.

———. 2007b. *Climate change, 2007. The physical science basis.* Geneva: Intergovernmental Panel on Climate Change. http://ipcc-wg1.ucar.edu/wg1/docs/WG1AR4_SPM_PlenaryApproved.pdf.

———. 2007b. *Working group II: Climate change impacts, adaptation, and vulnerability*. Geneva: Intergovernmental Panel on Climate Change. http://www.ipcc.wg2.org.

Isaacson, Walter. 2007. *Einstein: His life and universe*. New York: Simon and Schuster.

Jacoby, Susan. 2008. *The age of unreason*. New York: Pantheon.

Jasanoff, Sheila. 1995. *Science at the bar: Law, science, and technology in America*. Cambridge, Mass.: Harvard University Press.

———. 2005. Law's knowledge: Science for justice in legal settings. *American Journal of Public Health* 95 (S1): S49–58.

Johns Hopkins Medicine. 2005. Embryonic stem cells accrue genetic changes. http://www.hopkinsmedicine.org/Press_releases/2005/09_04_05.html.

Johnson, Mark. 1987. *The body in the mind: The bodily basis of meaning, imagination, and reason*. Chicago: University of Chicago Press.

———. 2006. A philosophical overview. *Daedalus* 135 (3): 46–54.

Johnson, Phillip E. 2000. *The wedge of truth: Splitting the foundations of naturalism*. Downers Grove, Ill.: Intervarsity.

John Templeton Foundation. 2008. Templeton prize. http://www.templeton.org/prizes/the_templeton_prize.

Kaiser, Jocelyn. 2008. Making clinical data widely available. *Science* 322:217–18.

Kass, Leon. 2001. In hearing before the Subcommittee on Health of the House Committee on Energy and Commerce, June 20. http://energycommerce.house.gov/reparchives/107/hearings/06202001Hearing291/Kass451.htm.

———. 2002. *Life, liberty, and the defense of dignity: The challenge for bioethics*. Pbk. ed. Washington, D.C.: AEI Press, 2004.

Kass, Leon, and James Q. Wilson. 1998. *The ethics of human cloning*. Washington, D.C.: AEI Press.

Kaye, David, and Jonathan Koehler. 1991. Can jurors understand probabilistic evidence? *Journal of the Royal Statistical Society* 154:79–80.

Keller, Evelyn Fox. 1985. *Reflections on gender and science*. New Haven, Conn.: Yale University Press.

Keller, Evelyn Fox, and Helen Longino, eds. 2006. *Feminism and science*. New York: Oxford University Press.

Kennedy, Donald. 2003. Forensic science: Oxymoron? *Science* 302:1625.

———. 2005. Silly season on the hill. *Science* 309:1301.

———. 2007. Turning the tables with Mary Jane. *Science* 316:661.

Kertzer, David I. 2006. *Prisoner of the Vatican*. New York: Mariner.

Kevles, Daniel J. 1990. Preface to *Science: The endless frontier*, by Vannevar Bush. 40th anniv. ed. Washington D.C.: National Science Foundation.

———. 1995. *The physicists: The history of a scientific community in modern America*. Cambridge, Mass.: Harvard University Press.

Kintsch, Eli. 2005. Researcher faces prison for fraud in NIH grant applications and papers. *Science* 307:1851.

Kitcher, Philip. 1982. *Abusing science: The case against creationism*. Cambridge, Mass.: MIT Press.

———. 1992. Authority, deference, and the role of individual reason. In *The social dimensions of science*, ed. Ernan McMullin. South Bend, Ind.: University of Notre Dame Press.

———. 1993. *The advancement of science*. New York: Oxford University Press.

————. 1994. Contrasting conceptions of social epistemology. In *Socializing epistemology*, ed. Frederick F. Schmitt. Lanham, Md.: Rowman and Littlefield.

————. 2001. *Science, truth, and democracy*. New York: Oxford University Press.

Klimanskaya, Irina, Young Chung, Sansy Becker, Shi-Jiang Lu, and Robert Lanza. 2006. Human embryonic stem cell lines derived from single blastomeres. *Nature* 444:481–85, 512.

Knorr-Cetina, Karen D., and Michael Mulkay, eds. 1983. *Science observed: Perspectives on the social study of science*. London: Sage.

Koehler, Jonathan J. 1997. Why DNA likelihood ratios should account for error. *Jurimetrics* 37:424–37.

————. 2001. The psychology of numbers in the courtroom: How to make DNA-match statistics seem impressive or insufficient. *Southern California Law Review* 74:1275.

Koertge, Noretta, ed. 1998. *A house built on sand: Exposing postmodernist myths about science*. New York: Oxford University Press.

Koestler, Arthur. 1959. *The sleepwalkers*. New York: Macmillan.

Kolata, Gina. 2007. Man who started stem cell wars may end it. *New York Times*, November 22. http://www.nytimes.com/2007/11/22/science/22stem.html.

Kolbert, Elizabeth. 2006. *Field notes from a catastrophe: A frontline report on climate change*. London: Bloomsbury.

Koshland, Daniel E., Jr. 1994. The DNA fingerprinting story (continued). *Science* 265:1015.

Krimsky, Sheldon. 1982. *Genetic alchemy: The social history of the recombinant DNA controversy*. Cambridge, Mass.: MIT Press.

Kuhn, Thomas S. 1962. *The structure of scientific revolutions*. 2nd ed., enlarged. Chicago: University of Chicago Press, 1970.

————. 1970. Reflections on my critics. In *Criticism and the growth of knowledge*, ed. Imre Lakatos and Alan Musgrave. New York: Cambridge University Press.

————. 1979. Metaphor in science. In *Metaphor and thought*, ed. Andrew Ortony. 2nd ed., enlarged. Cambridge: Cambridge University Press, 1993.

Labinger, Jay A. 2006. Organized skepticism, naïve methodism, and other –isms. *Foundations of Chemistry* 8:97–110.

Lakoff, George, and Mark Johnson. 1980. *Metaphors we live by*. Chicago: University of Chicago Press.

————. 1999. *Philosophy in the flesh: The embodied mind and its challenge to Western thought*. New York: Basic

Lander, Eric S., and Bruce Budowle. 1994. DNA fingerprinting dispute laid to rest. *Nature* 371:735–38.

Larsen, William J., Lawrence S. Sherman, S. Steven Potter, and William J. Scott. 1993. *Human embryology*. 3rd ed. Philadelphia: Churchill Livingstone, 2001.

Larson, Edward J. 1997. *Summer for the gods: The Scopes trial and America's continuing debate over science and religion*. Cambridge, Mass.: Harvard University Press.

Larson, Edward J., and Larry Witham. 1997. Scientists are still keeping the faith. *Nature* 386:435–36.

————. 1999. Scientists and religion in America. *Scientific American* 281 (3): 88–93.

Lasagna, Louis, and Sheila R. Shulman. 1999. Bendectin and the language of causation. In *Phantom risk: Scientific inference and the law*, ed. Kenneth R. Foster, David E. Bernstein, and Peter W. Huber. Cambridge, Mass.: MIT Press.

Latour, Bruno 1987. *Science in action*. Cambridge, Mass.: Harvard University Press.

Latour, Bruno, and Steve Woolgar. 1979. *Laboratory life: The construction of scientific facts*. 2nd rev. ed. Princeton, N.J.: Princeton University Press, 1986.

Lazer, David, and Michelle N. Meyer. 2004. DNA and the criminal justice system: Consensus and debate. In *DNA and the criminal justice system*, ed. David Lazer. Cambridge, Mass.: MIT Press.

Leshner, Alan I. 2008. Board's actions could put students at a disadvantage. *Houston Chronicle*, October 23. http://www.aaas.org/news/releases/2008/media/1024oped_texas/20081023houston_chron_oped.pdf.

Leuba, James H. 1916. *The belief in God and immortality: A psychological, anthropological, and statistical study*. Boston: Sherman, French.

Lewontin, Richard C., and Daniel L. Hartl. 1991. Population genetics in forensic DNA typing. *Science* 254:1745–50.

Lindberg, David C. 1992. *The beginnings of Western science*. Chicago: University of Chicago Press.

Linder, Douglas. 2000. The O. J. Simpson trial. *Jurist*. http://jurist.law.pitt.edu/trials10.htm.

Lindsey, Samuel, Ralph Hertwig, and Gerd Gigerenzer. 2003. Communicating statistical DNA evidence. *Jurimetrics* 43:147–63.

Lipton, P. 2005. The truth about science. *Philosophical Transactions of the Royal Society of London* 360 (1458): 1259–69.

Loftus, Elizabeth F., and Simon A. Cole. 2004. Contaminated evidence. *Science* 304:959.

Longino, Helen E. 1990. *Science as social knowledge: Values and objectivity in scientific inquiry*. Princeton, N.J.: Princeton University Press.

———. 2002. *The fate of knowledge*. Princeton, N.J.: Princeton University Press.

Luce, Henry R. 1941. The American century. *Life*, February 17, 61–65.

Lucretius. 1968. *The way things are [De rerum natura]*. Trans. Rolfe Humphries. Bloomington: Indiana University Press.

Lynch, Michael. 2003. God's signature: DNA profiling, the new gold standard in forensic science. *Endeavour* 27:93–97.

Lynch, Michael, Simon A. Cole, Ruth McNally, and Kathleen Jordan. 2008. *Truth machine: The contentious history of DNA fingerprinting*. Chicago: University of Chicago Press.

Lytle, Mark Hamilton. 2007. *The gentle subversive: Rachel Carson, Silent Spring, and the rise of the environmental movement*. New York: Oxford University Press.

Mackie, John L. 1977. *Ethics: Inventing right and wrong*. Harmondsworth, U.K.: Penguin.

Magnus, David, and Mildred K. Cho. 2005. Issues in oocyte donation for stem cell research. *Science* 308:1747–48.

Maitra, Anirban, Dan E. Arking, Narayan Shivapurkar, Morna Ikeda, Victor Stastny, Keyaunoosh Kassauei, Guoping Sui, et al. 2005. Genomic alterations in cultured human embryonic stem cells. *Nature Genetics* 37:1099–1103. http://www.nature.com/ng/journal/v37/n10/abs/ng1631.html.

Manabe, Syukuro, and Richard T. Wetherald. 1967. Thermal equilibrium of the atmosphere with a given distribution of relative humidity. *Journal of Atmospheric Research* 24:241–59.

Mann, Michael E., Raymond S. Bradley, and Malcolm K. Hughes. 1998. Global-scale temperature patterns and climate forcing over the past six centuries. *Nature* 392:779–87.

———. 1999. Northern Hemisphere temperatures during the past millennium: Inferences, uncertainties, and limitations. *Geophysical Research Letters* 26 (6): 759–62.

Margolis, Howard. 1991. Tycho's system and Galileo's dialogue. *Studies in the History and Philosophy of Science* 22:259–75.

McClory, Robert. 1995. *Turning point: The inside story of the papal birth control commission, and how Humanae Vitae changed the life of Patty Crowley and the future of the Church.* New York: Crossroad.

McGrath, Patrick J. 2002. *Scientists, business, and the state, 1890–1960.* Chapel Hill: University of North Carolina Press.

McMullin, Edward, ed. 2005. *The church and Galileo.* South Bend, Ind.: University of Notre Dame Press.

Meacham, Jon. 2007. Is God real? *Newsweek*, April 9. http://richarddawkins.net/article,825,The-God-Debate,Sam-Harris-Rick-Warren-Newsweek.

Meier, Barry. 2007. Narcotic maker guilty of deceit over marketing. *New York Times*, May 11. http://www.nytimes.com/2007/05/11/business/11drug.html.

Melnick, Ronald L. 2005. A *Daubert* motion: A legal strategy to exclude essential scientific evidence in toxic tort litigation. *American Journal of Public Health* 95 (S1): S30–34.

Merton, Robert K. 1949. *Social theory and social structure.* Enlarged ed. New York: Free Press, 1968.

———. 1973. *The sociology of science: Theoretical and empirical investigations.* Ed. N. W. Storer. Chicago: University of Chicago Press.

Miller, Arthur. 1954. *The crucible.* Pbk. ed. New York: Penguin, 2003.

Miller, Jon D. 2004. Public understanding of, and attitudes toward, scientific research: What we know and what we need to know. *Public Understanding of Science* 13:273–94.

Miller, Jon D., Rafael Pardo, and Fujio Niwa. 1997. *Public perceptions of science and technology.* Bilbao: Fundacion BBV.

Miller, Jon D., Eugenie C. Scott, and Shinji Okamoto. 2006. Public acceptance of evolution. *Science* 313:765–66.

Millikan, Robert A. 1923. Science and religion. *Science* 67:630.

Mnookin, Jennifer L. 2001a. Scripting expertise: The history of handwriting identification evidence and the judicial construction of reliability. *Virginia Law Review* 87:102–226.

———. 2001b. Fingerprint evidence in an age of DNA profiling. *Brooklyn Law Review* 67:14–71.

———. 2007. Idealizing science and demonizing experts: An intellectual history of expert evidence. *Villanova Law Review* 52: 763–801.

———. 2008. Expert evidence, partisanship, and epistemic competence. *Brooklyn Law Review* 73: 1009–34.

Molina, Mario J. 1996. Polar ozone depletion. *Angewandte Chemie International Edition* 35:1778–85.

Molina, Mario J., and F. Sherwood Rowland. 1974. Stratospheric sink for the chlorofluoromethanes: Chlorine atom catalyzed destruction of ozone. *Nature* 249:810–12.

Mooney, Chris, 2005. *The Republican war on science.* New York: Basic.

Morowitz, Harold J., and James S. Trefil. 1992. *The facts of life; Science and the abortion controversy.* New York: Oxford University Press.

Morton, Newton E. 1997. The forensic DNA endgame. *Jurimetrics Journal* 37:477–94.

Moyers, Bill. 2008. Religion and environment. http://www.pbs.org/moyers/moyersonamerica/green/environment.html.

Mueller, Christopher B. 1996. Introduction: O. J. Simpson and the criminal justice system on trial. *University of Colorado Law Review* 67:732.

Murphy, Priscilla Coit. 2005. *What a book can do: The publication and reception of "Silent Spring."* Amherst: University of Massachusetts Press.

NAS. 2000. *Who will do the science of the future? A symposium on careers of women in science.* Washington, D.C.: National Academies Press.

———. 2007a. *Beyond bias and barriers: Fulfilling the potential of women in academic science and engineering.* Washington, D.C.: National Academies Press.

———. 2007b. *Science, evolution, and creationism.* Washington, D.C.: National Academies Press.

National Academy of Sciences. *See* NAS.

National Institutes of Health. 2006. Conflict of interest information and resources. http://www.nih.gov/about/ethics_COI.htm.

National Oceanic and Atmospheric Administration, Office of Response and Restoration. 2008. http://response.restoration.noaa.gov/.

National Research Council. *See* NRC.

National Science Foundation. 2008. *Science and engineering indicators, 2008.* 2 vols. Arlington, Va.: National Science Foundation. http://nsf.gov/statistics/seind08/start.htm.

*Nature.* 2005. Climate of distrust. 436:1.

Neffe, Jürgen. 2007. *Einstein: A biography.* Trans. Shelley Frisch. New York: Farrar, Straus and Giroux.

Newport, Frank, and Maura Strausberg. 2001. American's belief in psychic and paranormal phenomena is up over last decade. Gallup News Service, June 8. http://www.gallup.com/poll/4483/Americans-Belief-Psychic-Paranormal-Phenomena-Over-Last-Decade.aspx.

*New York Tribune.* 1876. Science and religion once more. September 19. http://alepho.clarku.edu/huxley/comm/US/Trib.html.

Nisbet, Matthew C., and Chris Mooney. 2007a. Framing science. *Science* 316:56.

———. 2007b. The risks and advantages of framing science. *Science* 317:1169–70.

Noonan, John T., Jr. 1965. *Contraception: A history of its treatment by the Catholic theologians and canonists.* Enlarged ed. Cambridge, Mass.: Harvard University Press, 1986.

Nordhaus, William D., and Joseph Boyer. 2003. *Warming the world: Economic models of global warming.* Cambridge, Mass.: MIT Press.

Nowak, Rachel. 1994. Forensic DNA goes to court with O. J. *Science* 265:1352–54.

Nowotny, Helga. 2005. High- and low-cost realities for science and society. *Science* 308:1117–18.

NRC, Committee on DNA Technology in Forensic Science. 1992. *DNA technology in forensic science.* Washington, D.C.: National Academies Press.

———. 1996. *DNA technology in forensic science: An update.* Washington, D.C.: National Academies Press.

―――, Commission on Life Sciences. 2001. *Evaluating chemical and other agent exposures for reproductive and developmental toxicity.* Washington, D.C.: National Academies Press.

―――, Science, Technology, and Law Panel. 2002. *The age of expert testimony: Science in the courtroom: Report of a workshop.* Washington, D.C.: National Academies Press.

―――, Board on Life Sciences. 2005. *Guidelines for human embryonic stem cell research.* Washington, D.C.: National Academies Press.

―――, Committee to Review the OMB Risk Assessment Bulletin. 2007. *Scientific review of the proposed risk assessment bulletin from the Office of Management and Budget.* Washington, D.C.: National Academies Press.

Patton, Stephen M. 1990. DNA fingerprinting: The *Castro* case. *Harvard Journal of Law and Technology* 3:223–40.

Patzig, Günther. 2002. Can moral norms be rationally justified? *Angewandte Chemie International Edition* 41 (18): 3353–58.

Pew Center on Global Climate Change. 2008a. Greenhouse gas emission targets. http://www.pewclimate.org/what_s_being_done/in_the_states/emission stargets_map.cfm.

―――. 2008b. Business. http://www.pewclimate.org/business.

Pew Global Attitudes Project. 2007. *World publics welcome global trade—but not immigration.* Washington, D.C.: Pew Research Center. http://pewglobal.org/reports/pdf/258.pdf .

Pew Research Center for the People and the Press. 2006. Online papers modestly boost newspaper readership: Maturing internet news audience broader than deep. http://people-press.org/reports/display.php3?PageID=1064.

Philipkoski, Kristen. 2003. Bioethics shuffle ignites outcry. *Wired*, March 2. http://www.wired.com/news/medtech/0,1286,62494,00.html.

Phillips, Kristen E. 2007. Seas could rise dramatically in rapid ice melt. *LiveScience*, April 4. http://www.livescience.com/environment/070404_GW_ice.html.

Pickering, Andrew, ed. 1992. *Science as practice and culture.* Chicago: University of Chicago Press.

Pius XI. 1930. *Casti Connubii.* http://www.vatican.va/holy_father/pius_xi/encyclicals/documents/hf_p-xi_enc_31121930_casti-connubii_en.html.

Polanyi, Michael. 1945a. The autonomy of science. *Scientific Monthly* 60:141–50.

―――. 1945b.The growth of thought in society. *Economica*, November, 428–56.

―――. 1946. *Science, faith, and society.* Chicago: University of Chicago Press.

―――. 1958. *Personal knowledge: Toward a post-critical philosophy.* Chicago: University of Chicago Press.

―――. 1962. The republic of science: Its political and economic theory. *Minerva* 1:54–74.

―――. 1966. *The tacit dimension.* Repr. ed. Gloucester, Mass.: Peter Smith, 1983.

Polinghorne, John. 1998. *Belief in God in an age of science.* New Haven, Conn.: Yale University Press.

*Popular Science Monthly.* 1873. Vol. 2, November–April. New York: D. Appleton.

Prentice, David A. 2005. Live patients and dead mice. *Christianity Today*, October. http://www.christianitytoday.com/ct/2005/october/24.71.html

Prentice, David A., and Gene Tarne. 2007. Treating diseases with adult stem cells. *Science* 315:328.

Price, Don K. 1978. Endless frontier or bureaucratic morass? In *Limits of scientific inquiry*, ed. Gerald Holton and Robert S. Morison. New York: Norton.

Project for Excellence in Journalism. 2007. The state of the news media, 2007. http://www.stateofthemedia.org/2007/narrative_radio_talk_radio.asp?cat=8&media=9.

Rabb, Theodore K. 2006. *The last days of the Renaissance*. New York: Basic.

Raymo, Chet. 1998. *Skeptics and true believers: The exhilarating connection between science and religion*. New York: Walker.

Raz, Joseph. 1990. Introduction. In *Authority*, ed. Joseph Raz. Oxford: Blackwell.

RealClimate. 2005. Scientists respond to Barton. http://www.realclimate.org/index.php/archives/2005/07/barton-and-the-hockey-stick.

Reichler, Thomas, and Junsu Kim. 2008. How well do coupled models simulate today's climate? *Bulletin of the American Meteorological Society* 89 (3): 303–11.

Reid, Thomas. 1764. *An inquiry into the human mind*. Ed. Timothy Duggan. Chicago, University of Chicago Press, 1970.

Revkin, Andrew C. 2005. Bush aide softened greenhouse gas links to global warming. *New York Times*, June 8. http://www.nytimes.com/2005/06/08/politics/08climate.html.

———. 2006. Climate expert says NASA tried to silence him. *New York Times*, January 29. http://www.nytimes.com/2006/01/29/science/earth/29climate.html.

Rhodes, Richard. 1986. *The making of the atomic bomb*. New York: Simon and Schuster.

Robinson, Emily. 2007. Exxon exposed. *Catalyst* 6:2–4.

Rorty, Richard. 1989. Science as solidarity. In *Dismantling truth: Reality in the postmodern world*, ed. Hilary Lawson and Lisa Appignansei. New York: St. Martin's Press.

Rothman, Kenneth J., and Sander Greenland. 2005. Causation and causal inference in epidemiology. *American Journal of Public Health* 95 (S1): S144–50.

Roughgarden, Joan. 2006. *Evolution and Christian faith: Reflections of an evolutionary biologist*. Washington, D.C.: Island.

Rowland, F. Sherwood. 1996. Stratospheric ozone depletion by chlorofluorocarbons. *Angewandte Chemie International Edition* 35:1786–98.

Rowley, Janet D., Elizabeth Blackburn, Michael S. Gazzaniga, and Daniel W. Webster. 2002. Harmful moratorium on stem cell research. *Science* 297:1957.

Rubin, Joel, and Richard Winton. 2008. LAPD blames faulty fingerprint analysis for erroneous accusations. *Los Angeles Times*, October 17. http://www.latimes.com/news/local/la-me-fingerprints17-2008oct17,0,6045556.story.

Ruse, Michael. 2005. *The evolution-creation struggle*. Cambridge, Mass.: Harvard University Press.

Saks, Michael J., and David L. Faigman. 2005. Expert evidence after *Daubert*. *Annual Review of Law and Social Science* 1:105–30.

Saks, Michael J., and Jonathan J. Koehler. 2005. The coming paradigm shift in forensic identification science. *Science* 309:892–95.

Sallo, Todd. 1999. Brother can you paradigm? *National CrossTalk*. http://www.highereducation.org/crosstalk/ct1099/voices1099-sallo.shtml.

Sandoval, William. 2005. Understanding students' practical epistemologies and their influence on learning through inquiry. *Science Education* 89 (4): 634–56.

Schardein, James L. 2000. *Chemically induced birth defects*. 3rd ed. New York: Marcel Dekker.

Scheideler, Britta. 2000. The scientist as moral authority: Albert Einstein between elitism and democracy, 1914–1933. *Historical Studies in the Physical and Biological Sciences* 31:319–46.

Schreiber, Stuart. 2005. Small molecules: The missing link in the central dogma. *Nature Chemical Biology* 1:64–66.

Schuck, Peter H. 1986. *Agent Orange on trial: Mass toxic disasters in the courts.* Enlarged ed. Cambridge, Mass.: Harvard University Press, 1987.

Schulz, William G. 2006. Judging science. *Chemical and Engineering News*, February 27, 36–39.

Schweber, Silvan S. 2000. *In the shadow of the bomb.* Princeton, N.J.: Princeton University Press.

Seligman, Adam B. 2000. *Modernity's wager: Authority, the self, and transcendence.* Princeton, N.J.: Princeton University Press.

Shannon, Thomas A., and Allan B. Wolter. 1990. Reflections on the moral status of the pre-embryo. *Theological Studies* 51 (4): 603–26.

Shapin, Steven. 1994. *A social history of truth: Civility and science in seventeenth-century England.* Chicago: University of Chicago Press.

———. 1995. Trust, honesty, and the authority of science. In *Society's choices: Social and ethical decision making in biomedicine*, ed. Ruth Ellen Bulger, Elizabeth Meyer Bobby, and Harvey V. Fineberg. Washington, D.C.: National Academies Press.

———. 2008. *The scientific life: A moral history of a late modern vocation.* Chicago: University of Chicago Press.

Shea, William R., and Mariano Artigas. 2003. *Galileo in Rome: The rise and fall of a troublesome genius.* Oxford: Oxford University Press.

Shreeve, James. 2004. *The genome war: How Craig Ventner tried to capture the code of life and save the world.* New York: Knopf.

Siegenthaler, Urs, Thomas F. Stocker, Eric Monnin, Dieter Lüthi, Jakob Schwander, Bernhard Stauffer, Dominique Raynaud, et al. 2005. Stable carbon cycle–climate relationship during the late Pleistocene. *Science* 310:1313–17.

Simon, Yves R. 1962. *A general theory of authority.* South Bend, Ind.: University of Notre Dame Press.

Simpson, Joe Leigh. 2006. Medicine: Blastomeres and stem cells. *Nature* 444: 432–35.

Smith, Shane, William Neaves, and Steven Teitelbaum. 2006. Adult stem cell treatments for diseases? *Science* 313:439.

Snow, C. P. 1959. *The two cultures.* Pbk. ed. Introd. by Stefan Collini. Cambridge: Cambridge University Press, 1993.

Sobel, Dava. 2000. *Galileo's daughter.* New York: Putnam.

Solomon, Susan. 1988. The mystery of the Antarctic ozone "hole." *Reviews of Geophysics* 26:131–48.

Sonnert, Gerhardt, and Gerald Holton. 1999. A vision of Jeffersonian science. *Issues in Science and Technology* 16 (1): 61–65.

———. 2002. *Ivory bridges: Connecting science and society.* Cambridge, Mass.: MIT Press.

Specker, Steven. 2007. Electricity technology in a carbon-constrained future. http://mydocs.epri.com/docs/CorporateDocuments/Newsroom/EPRIUSElectSectorCO2Impacts_021507.pdf

Stark, Rodney. 2003. *For the glory of God: How monotheism led to reformations,* ѕсіenсе, witch hunts, and the end of slavery. Princeton, N.J.: Princeton University Press.

Stark, Rodney, and William Sims Bainbridge. 1987. *A theory of religion*. Republished ed. New Brunswick, N.J.: Rutgers University Press, 1996.

Stokes, Donald E. 1997. *Pasteur's quadrant: Basic science and technological innovation*. Washington, D.C.: Brookings Institution Press.

Stokstad, Erik. 2003. EPA report takes heat for climate change edits. *Science* 300: 2013.

Stolarski, Richard S., and Ralph J. Cicerone. 1974. Stratospheric chlorine—possible sink for ozone. *Canadian Journal of Chemistry* 52:1610–15.

Stolberg, Sheryl Gay. 2005. Senate leader criticized and praised for stem cell shift. *New York Times*, July 30. http://www.nytimes.com/2005/07/30/politics/30 stem.html.

Sulston, John, and Georgina Ferry. 2002. *The common thread: A story of science, politics, ethics, and the human genome*. Washington, D.C.: National Academies Press.

Sunstein, Cass R. 2007. *Worst-case scenarios*. Cambridge, Mass.: Harvard University Press.

Takahashi, Kazutoshi, Koji Tanabe, Mari Ohnuki, Megumi Narita, Tomoko Ichisaka, Kiichiro Tomoda, and Shinya Yamanaka. 2007. Induction of pluripotent stem cells from adult human fibroblasts by defined factors. *Cell* 131:1–12. http://images.cell.com/images/Edimages/Cell/IEPs/3661.pdf.

Talbot, Margaret. 2005. Darwin in the dock. *New Yorker*, December 5, 66–77.

Teixeira, Ruy. 2008. What the public really wants on science. *Center for American Progress*, July 16. http://www.americanprogress.org/issues/2008/07/wtprw _science.html.

Thompson, William C. 1996. DNA evidence in the O. J. Simpson trial. *University of Colorado Law Review* 67:826.

———. 1997. The National Research Council's second report on forensic DNA evidence: A critique. *Jurimetrics* 37:405–24.

Thornton, John, and Joseph Peterson. 2002. The general assumptions and rationale of forensic identification. In *Modern scientific evidence: The law and science of expert testimony*, ed. David L. Faigman, David H. Kaye, and Michael J. Saks. 2nd ed. Minneapolis: West/Thompson, 2005.

Thorpe, Charles. 2001. Science against modernism: The relevance of the social theory of Michael Polanyi. *The British Journal of Sociology* 52 (1): 19–35.

Timmerman, Luke, and David Heath. 2005. Drug researchers leak secrets to Wall Street. *Seattle Times*, August 7. http://seattletimes.nwsource.com/html/bus inesstechnology/drugsecrets1.html.

Toobin, Jeffrey. 2007. The CSI effect. *New Yorker*, May 7, 30–37.

Tyndall Centre. 2008. John Tyndall, FRS, DCL, LLD. http://www.tyndall.ac.uk/gen eral/history/john_tyndall_biography.shtml.

Union of Concerned Scientists. 2007. *Smoke, mirrors and hot air: How ExxonMobil uses big tobacco's tactics to manufacture uncertainty on climate science*. Cambridge, Mass.: Union of Concerned Scientists. http://www.ucsusa.org/assets /documents/global_warming/exxon_report.pdf.

United States Climate Action Partnership. 2008. Home page. http://www.us-cap .org.

U.S. Commerce Department. 2007. Frequently asked questions regarding the

Department of Commerce public communications policy. http://www.com merce.gov/opa/press/Secretary_Gutierrez/2007_Releases/March/29_FAQ .pdf.

U.S. Conference of Catholic Bishops. n.d. Respect for unborn human life: The church's constant teaching. http://www.usccb.org/prolife/constantchurch teaching.shtml.

———. 2001. Ethical and religious directives for Catholic health care services. http://www.usccb.org/bishops/directives.shtml#partfour.

U.S. Congress. House. 2006. *Hearings before the subcommittee on oversight and investigations of the Committee on Energy and Commerce.* 109th Congress, 2nd sess., July 19 and July 27, 2006.

———. Senate. 1997. *Byrd-Hagel resolution.* S Res 98. 105th Cong., 1st sess. http: //www.nationalcenter.org/KyotoSenate.html.

U.S. *Congressional Record.* U.S. Senate. 2005. *Global warming debate,* January 4, 518.

U.S. Environmental Protection Agency. 2008. The phaseout of methyl bromide. http://www.epa.gov/spdpublc/mbr.

U.S. Food and Drug Administration. 2001. Oxycontin: Questions and answers. http://www.fda.gov/cder/drug/infopage/oxycontin/oxycontin-qa.htm.

van Fraassen, Bas C. 1980. *The scientific image.* Oxford: Clarendon.

Veblen, Thorstein. 1919. *The place of science in modern civilization, and other essays.* New York: B. W. Heubsch.

Virginia Commonwealth University, Center for Public Policy. 2001. Americans welcome scientific advancements with caution. http://www.news.vcu.edu/ news.aspx?v=detail&nid=751.

Vogel, Gretchen, and Constance Holden. 2007. Developmental biology: Field leaps forward with new stem cell advances. *Science* 318:1224–25.

Walters, Ronald G. 1997. Introduction: Uncertainty, science, and reform in twentieth-century America. In *Scientific authority and twentieth-century America,* ed. Ronald G. Walters. Pbk. ed. Baltimore: Johns Hopkins University Press, 2001.

Walton, Douglas. 1997. *Appeal to expert opinion: Arguments from authority.* University Park: Pennsylvania State University Press.

Wambaugh, Joseph. 1989. *The blooding.* New York: Morrow.

Weart, Spencer R. 2003. *The discovery of global warming.* Cambridge, Mass.: Harvard University Press.

Weber, Max. 1915. *The theory of social and economic organization.* Trans. A. M. Henderson and T. Parsons. New York. Free Press, 1947.

———. 1918. *The vocation lectures: Science as vocation, politics as vocation.* Trans. David S. Owen, Tracy B. Strong, and Rodney Livingstone. Indianapolis: Hackett, 2004.

———. 1922. *The sociology of religion.* Trans. Ephraim Fischoff. Boston: Beacon, 1993.

Weinberg, Steven. 1994. *Dreams of a final theory: The scientist's search for the ultimate laws of nature.* New York: Vintage.

Weinshiboum, Richard. 2003. Inheritance and drug response. *New England Journal of Medicine* 348 (6): 529–37.

Weinstein, Bernard. 1991. Mitogenesis is only one factor in carcinogenesis. *Science* 251:387–88.

White, Andrew Dickson. 1896. *A history of the warfare of science with theology in Christendom.* New York: George Braziller, 1955. http://abob.libs.uga.edu/ bobk/whitewtc.html.

White House. 2001. Remarks by the president on stem cell research. http://www
.whitehouse.gov/news/releases/2001/08/20010809–2.html.

White House Executive Order 2009. http://www.whitehouse.gov/the_press_office/
Removing-Barriers-to-Responsible-Scientific-Research-Involving-Human
-Stem-Cells/

Wikipedia. 2008. Wikipedia: About. http://en.wikipedia.org/wiki/Wikipedia:About.

Wills, Christopher. 1992. Forensic DNA typing. *Science* 255:1051–55.

Wills, Garry. 2000. *Papal sin: Structures of deceit.* New York: Doubleday.

Wilson, Edward O. 1999. *Consilience: The unity of knowledge.* New York: Vintage.

———. 2006. *The creation: An appeal to save life on Earth.* New York: Norton.

Wilson, James F., Michael E. Weale, Alice C. Smith, Fiona Gratrix, Benjamin
Fletcher, Mark G. Thomas, Neil Bradman, et al. 2001. Population genetic
structure of variable drug response. *Nature Genetics* 29:265–69.

Winston, Robert Lord. 2005. Should we trust the scientists? Presented June 20 at
Gresham College, London.http://www.gresham.ac.uk/event.asp?PageId=39
&EventId=347.

Wippel, John F. 1998. Siger of Brabant. In *Routledge Encyclopedia of Philosophy,* ed.
Edward Craig. London: Routledge.

Wohn, D. Yvette, and Dennis Normile. 2006. "Korean cloning scandal: Prosecu-
tors allege elaborate deception and missing funds." *Science* 312:980–81.

Wojdacz, Mariah. n.d. Til divorce do us part: What's the real status of marriage
in America? http://www.legalzoom.com/articles/article_content/article13746
.html.

Wolpert, Lewis. 2005. Is science dangerous? *Philosophical Transactions of the Royal
Society of London* 360 (1458): 1253–58.

———. 2007. *Six impossible things before breakfast: The evolutionary origins of belief.*
New York: Norton.

Yankelovich, Daniel. 2003. Winning greater influence for science. *Issues in Science
and Technology* 19 (4). http://www.issues.org/19.4/yankelovich.html.

Yu, Junying, Maxim A. Vodyanik, Kim Smuga-Otto, Jessica Antosiewicz-Bourget,
Jennifer L. Frane, Shulan Tian, Jeff Nie, et al. 2007. Induced pluripotent
stem cell lines derived from human somatic cells. *Science* 318:1917–20.

Zarin, Deborah A., and Tony Tse. 2008. Moving toward transparency of clinical trials. *Science* 319:1340–42.

Ziman, John M. 1984. *An introduction to science studies.* Cambridge: Cambridge
University Press.

———. 2000. Commentary on *The republic of science:* Its political and economic
theory. *Minerva* 38 (1): 21–25.

LEGAL REFERENCES

*Agent Orange Product Liability Litigation.* 1985. 611, F. Supp. 1223 (E.D. N.Y.).
*Daubert v. Merrell Dow Pharmaceuticals Inc.* 1993. 92–102, 509 U.S. 579.
*Epperson v. Arkansas.* 1968. 393 U.S. 97.
*Federal Rules of Evidence.* 2007. http://www.law.cornell.edu/rules/fre.
*Folkes, L. v. Chadd, 3 Doug.* 1782. 157.
*Frye v. United States.* 1923. 293 F. 1013, 1014 (D.C. Cir.).
*General Electric Company v. Joiner.* 1997. 522 U.S. 136, 118, S. Ct. 512.

*Harris v. McRae.* 1980. 448 U.S. 297.

*Kitzmiller et al. v. Dover Area School District et al.* 2005. Mid. Dist. Pa., 04cv2688.

*Kumho Tire Company v. Carmichael.* 1999. 526 U.S. 137, 119, S. Ct. 1167.

*McLean v. Arkansas Board of Education.* 1982. 529 F. Supp. 1255, 50 U.S. Law Week 2412.

*Moon v. State.* 1921. 198 P. 288 (Ariz.).

*People v. Castro.* 1989. 545 N.Y. 2d 985 S. Ct.

*Richardson-Merrell, Inc., "Bendectin" Products Liability Litigation, In re.* 1985.

*United States v. Havvard.* 2000. 117 F. Supp. 2d 848 (D.C. Ind.).

*United States v. Mitchell.* 1999. Crim. No. 96–407 (E.D. Pa.).

*United States v. Plaza.* 2002. Crim. No. 98–362 (E.D. Pa.).

# INDEX